NextGen Network Synchronization

Dhiman Deb Chowdhury

NextGen Network
Synchronization

 Springer

Dhiman Deb Chowdhury
San Jose, CA, USA

ISBN 978-3-030-71181-8   ISBN 978-3-030-71179-5   (eBook)
https://doi.org/10.1007/978-3-030-71179-5

This Springer imprint is published by the registered company Springer Nature Switzerland AG
The registered company address is: Gewerbestrasse 11, 6330 Cham, Switzerland

*This Book is dedicated to my father,
Dr. Dhirendra Chandra Deb Choudhury
(1932–2019),with love. Your humble
beginning to pinnacle of success sets a great
example of what is possible when passion,
action, discipline, and perseverance are put
together. Your banyan reflection has been and
remains the guiding force for us—I miss
you, Dad.*

Preface

Overlooking the vast waterways of Surma plains in the northeastern part of the Bengal delta, I observed a natural phenomenon in my childhood that to date is vivid in my mind. As migratory birds flock over the waterways during the dawn and twilight, they appear to maintain a natural symmetry of forming a sign wave across the distant horizon. The waves of flocks are seemingly phase aligned.

We rarely think about this naturally occurring phenomenon when it comes to the design of network infrastructure, yet it is something de facto in how communication devices work in synchronicity. Whether a network is homogenous or heterogeneous, imperatives of synchronicity cannot be overlooked, nor can it be ignored. In every device in a network from computers to routers, an inherent synchronization is at work by design thanks to the advent of oscillators for local reference and NTP for distributed synchrony. For many of us, looking under the hood is not something we do often when it comes to designing the network. However, given the increased applications of time-sensitive transport in different industry verticals, synchrony and, for that matter, high precision synchronicity, can no longer be assumed or overlooked.

In some applications such as industrial automation, power utilities, and 5G transport, high-precision synchrony is a must, and without a proper sync plane design network will suffer significant performance issues. Whereas industrial automations are concerned such issues may lead in catastrophic failure resulting in safety hazards.

The primary objective of this book is therefore to teach the reader about the science of synchronicity, technologies, devices, parameters, and guidelines for the proper design of sync for the nextgen networks. The book blends theory, guidelines from applicable standards, and practices to enable readership to gradually learn, and in process, attain mastery of sync plane design for nextgen networks in different industry verticals that include telecom clouds and 4G/5G fronthaul, enterprise 5G such as CBRS, CATV, datacenter, and industrial networks. There are two distinct sources of reference clock for synchrony: atomic clock and clock derived through GNSS (Global Navigation Satellite System). Both of these technologies are explored in the book with an in-depth discussion on GNSS-based reference clock systems due to its common use as PRTC (Primary Reference Time Clock) for synchrony. This discussion is followed by two distinct time distribution methods to achieve synchrony across the network: frequency-based and packet-based. Later, the concept learned is applied to modern network infrastructure with a specific discussion related to sync plane design of:

- Telecommunication Network Infrastructure
- SmartGrid
- Data Center and MSO (CATV) Network Infrastructure
- Industrial Networks

A detailed design book, resplendent with discussions on primary reference clock systems, GNSS devices, timing parameters, protocols, and overall sync plane design of nextgen networks, speaks volumes. In fact, each chapter could be extended to a book, yet the author made every attempt to furnish as many details as possible to provide a comprehensive purview of the subject in each chapter. It is, however, recommended that readers who are interested to learn further should study appropriate industry standards.

Author's Note

This book attempts to provide comprehensive, accurate, and timely information up to the date of publication. While many of the industry standards presented in the book are final, some may be in the process of approval, and thus, it is advisable for the readership to further explore those industry standards for their applicability. Please excuse some of the blatant assumptions that users are ready to implement some of the advanced technologies—on rare occasions, it will take some time before they can be implemented. Please also excuse my personal biases, which may have crept into the text in a few sync planes examples. I simply presented some of the products best known to me to present the type of product that is used in those scenarios to design sync planes. This is by no means to promote any product or technologies.

San Jose, CA, USA   Dhiman Deb Chowdhury

Contents

Chapter 1
Fundamentals of Time Synchronization

1.1 Introduction

Synchronization is often taken for granted yet the fundamental conduits of an optimized digital communications infrastructure. As ubiquitous connectivity transforming industries blending physical and digital world, our desire for enhanced experience and business insights of data gathered at physical endpoints is pushing the boundaries of service through real-time and near real-time processing of data. This demand requires accuracy and stability of signals or transport that passes through various devices in a given network. Time synchronization is thus a key imperative of modern communications networks. In this chapter, we will learn the concept of synchronization and types of synchronization that are applicable to data transport or communication networks and more about the device element that measures passage of time, "the clock." We will advance our understanding by exploring further on clock and types of clock including how time is defined and measured.

In proceeding chapters, we will explore how clock frequency references are distributed across the network to keep all devices in sync and thus, help in the optimization of network transport. Henceforth, fundamentals presented in this chapter will serve as the basis on which to extend our understanding on the importance of time and methods of its distribution across networks.

1.2 Synchronization

In our increasingly connected digital economy, the communication infrastructure we rely on are built on synchronization of clocks. It is about a common notion of time that all nodes in a communication infrastructure must agree upon. This common notion of time could be a fraction of a second to a billionth of a second depending

D. D. Chowdhury, *NextGen Network Synchronization*, https://doi.org/10.1007/978-3-030-71179-5_1

upon applications and network scenarios. For example, radio unit in a cell tower may require couple of nanosecond-level accuracy while a laptop browsing internet can tolerate at tens of milliseconds level accuracy.

In practice, all network and computing devices in a communication infrastructure have its own clock to keep track of current time. These clocks can easily drift to a second per day and thus accumulating significant error over time. In addition, even if all devices synchronized at the beginning, they may not remain synchronized since clocks at different devices tick at different rates. Thus, it is imperative for all systems or devices in a given network to periodically synchronize to a precise clock or time depending upon time synchronization requirements for the network and applications.

1.2.1 Synchronization Types

The technique that coordinates these clocks to a common notion of time is known as clock or time synchronization. ITU-T recommendation G. 8260 defines three types of synchronizations:

- Frequency Synchronization (a.k.a. syntonization): This is a most common form of synchronization and generally used in point to point communication link. As illustrated in Fig. 1.1, two systems A and B get a common reference signal and evolve at the same rate within given accuracy and stability but some of the significant aspects of the signal, e.g., phase, are not aligned in time [1].
- Phase Synchronization: An important characteristic of a signal is the phase. It specifies the location of a point within a wave cycle of a repetitive waveform. As illustrated in Fig. 1.2, two systems A and B get a common reference signal that evolves at the same rate and their phases are aligned in time but are not traceable to a reference clock.
- Time Synchronization: It is the third component of a traceable common reference time that can be assigned to phase synchronization. In other words, time synchronization combines frequency and phase synchronization with a common traceable reference time across the networks. As depicted in Fig. 1.3, systems A and B are phase synchronized with a known traceable reference clock (e.g., UTC).

- Examples of common reference times are [2]:

  - UTC
  - International Atomic Time (TAI)
  - UTC + offset (e.g., local time)
  - Global Positioning System (GPS)
  - PTP, NTP, etc.
  - local arbitrary time

It is to be noted that distributing time synchronization is one way of achieving phase synchronization.

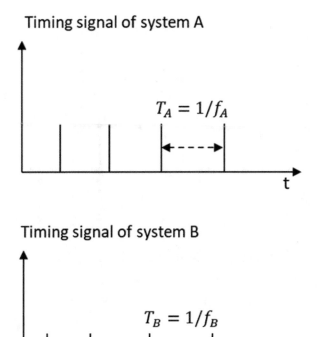

Fig. 1.1 Diagrammatical representation of frequency synchronization

1.3   Clock

The word "clock" here refers to a device that measures either a point in time (time of day) or the passage of time (time interval). It comprises of a stable oscillator and a counter. An oscillator is composed of a crystal to which an electrical charge is applied. This charge causes the crystal to vibrate and oscillate at a particular frequency, producing an electrical signal with a given frequency in the process. This frequency is commonly used to keep track of time. For example, wristwatches are used in digital integrated circuits to provide a stable clock signal and also used to stabilize frequencies for radio transmitters and receivers. For a precise clock implementation, a quartz oscillator or atomic oscillator may be used. The quartz oscillator uses quartz crystal made of mineral composed of silicon and oxygen atoms. When a voltage source is applied to quartz crystal, these materials react and generate an electric charge in response to applied mechanical stress. This characteristic is known as piezoelectric effect. This phenomenon was first discovered in 1880 by two French scientists and brothers, Jacques and Pierre Curie. While

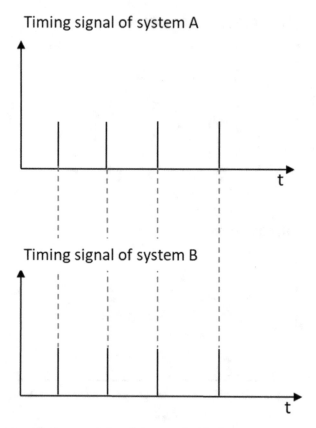

Fig. 1.2 Diagrammatical representation of phase synchronization

experimenting with crystals, they found that applying mechanical pressure to specific crystals like quartz released an electrical charge. They called this the piezoelectric effect. For 30 years, piezoelectricity remained apart of laboratory experiment. It was not until World War I when piezoelectric effect was first found its practical application in sonar.

The piezoelectric effect is the byproduct of electromechanical interactions. Figure 1.4 illustrates the electromechanical effect on quartz crystal of a crystal oscillator leading to piezoelectricity. As the voltage applied, a mechanical tension or compression is developed in the crystal. It pulls positive and negative ions present in the crystals away from each other, creating an energy gradient across the crystal allowing an electric current to flow.

The piezoelectric crystals such as quartz are cut and fabricated in the form of wafer whose physical size and thickness are tightly controlled since it affects the final or fundamental frequency of oscillations or "characteristic frequency." This crystal characteristic frequency is inversely proportional to physical thickness of the crystal wafer.

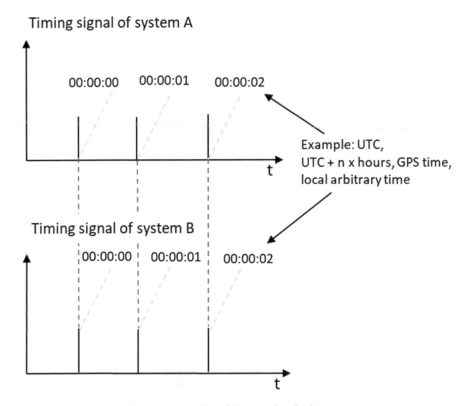

Fig. 1.3 The diagrammatical representation of time synchronization

1.3.1   Offset, SKEW, and Drift

The offset, skew, and drifts are different sources of error in a timing device. IETF RFC 2330 defines offset as the difference between time reported by the clock and the "true" time as defined by the UTC (Universal Coordinated Time). We will discuss about UTC later in this chapter. To our understanding, let's consider clock reported time is Tc and true time reported UTC is Tt. In this case, the clock offset is Tc-Tt. If the clock offset value is close to zero, it is considered accurate. The term of accuracy also includes a notion of frequency of the clock. A clock's frequency is its cycles per true second defined by UTC. The skew in this case would be the difference between clock's frequency and true time frequency. Drift can be described as the change in accuracy of a given frequency over environmental changes such as temperature, humidity, or pressure, or simply over long periods of time. It is measured in parts per million (ppm).

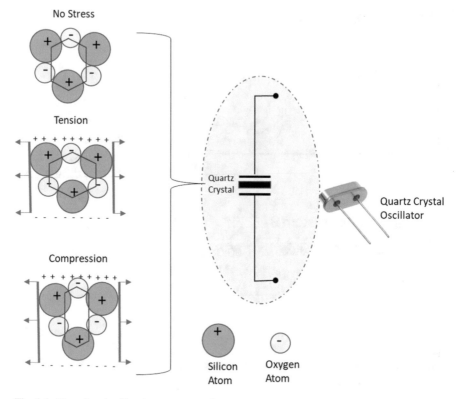

Fig. 1.4 Piezoelectric effect in quartz crystal

1.3.2 Crystal Oscillator as Clock Source

In a clock implementation, the fundamental frequency of crystal is divided, and resultant pulses are counted. The crystal used in desktop computer clock application generally produces 32.768 kHz. A 16bit counter acting as count down divider can divide this frequency to a 1-Hz output (2^{15} = 32,768). The 1-Hz counter output which occurs once every 32,768 pulses is then used to drive the clock display or internal clock timestamp. A typical quartz crystal oscillator can produce 10s of KHz to well over 100 MHz. Figure 1.5 depicts a typical CMOS clock oscillator circuit. Most computer today includes a CMOS clock oscillator with battery backup.

The figure shows a parallel-resonant[1] configuration of a basic CMOS[2] oscillator. The product can be made using CMOS inverters (I1 & I2) to achieve required

---

[1] Parallel resonance means it is a parallel circuit which will be operating at below, above, or at resonant (low impedance at a certain frequency) depending upon frequency and component values.

[2] CMOS or complementary metal-oxide semiconductor is a technology that is used to produce integrated circuit.

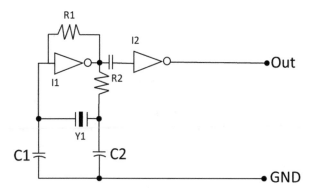

Fig. 1.5 Typical CMOS clock oscillator circuit

amplitude. The inverter I1 is acting as a "Schmitt Trigger"[3] in inverse configuration. The crystal Y1 will provide oscillation in series resonance frequency. R1 is a feedback resistor to the CMOS inverter. I2 is acting as booster to provide sufficient output for the load. The circuitry provides square wave output for which maximum output frequency depends on the switching characteristics of the CMOS inverters. The changes in capacitors and resistors values will also change the output frequency. This output frequency is used by CPUs and microcontrollers as clock frequency.

1.3.3   Crystal Oscillators Types

Depending upon compensation techniques used to achieve higher precision and accuracy, crystals can be divided into four categories and those are:

- *Uncompensated crystal oscillator (XO)*: The XO provides excellent frequency stability without temperature compensation and frequency control. It provides a clock output at a specified frequency and signal format (e.g., CMOS, LVDS, LVPECL).
- *Temperature controlled oscillator (TCXO)*: A TCXO is specially designed to operate in high temperature environment. It has temperature-sensitive reactance circuit in its oscillation loop to compensate frequency-temperature characteristics inherent to the crystal unit. Generally, the oscillation frequency of crystal oscillator like XO fluctuates as temperature rises resulting unstable frequency output. In contrast, TCXO provides stable output frequency in such high temperature environment.

---

[3] Schmitt trigger is a digital logic that can perform arithmetic and logical operations. It is a comparator circuit meaning it compares two voltages or currents and outputs a digital signal indicating which is larger.

- *Oven controlled oscillator (OCXO)*: Both TCXO and OCXO oscillation frequency do not vary at high temperatures or when the temperature fluctuates, but OCXO power consumption is relatively higher than TCXO. Unlike TCXO which uses temperature compensated circuit, it is placed in an oven that is pre-heated to a higher temperature and thus external temperature variations do not impact output frequency. It is to be noted that OCXO provides far better frequency stability and performance than TCXO and is suitable for applications where precise frequency is needed, e.g., telecom equipment and military applications.
- *Voltage controlled crystal oscillator (VCXO)*: It is a XO whose frequency is determined by a crystal, but it can be adjusted or tuned by applying control voltage to the input.

1.4   Atomic Oscillator

Crystal oscillator is subject to frequency drift that causes clocks to lose accuracy over time. In contrary, atomic oscillator nearly has no frequency drift. It uses the quantized energy levels in atoms or molecules as the source of its resonance. According to quantum mechanics, atom has discrete energy levels. A magnetic field at a particular frequency can either boost or drop the energy level of an atom.[4] According to NIST, the resonance frequency or better known as photonic frequency "$f0$" of an atomic oscillator is the difference between two energy levels divided by Planck's constant, "h":

$$f0 = \frac{E2 - E1}{h}$$

[where $E2$ and $E1$ are the upper and lower energy states].

Therefore, a transition occurring at a well-defined frequency can be denoted as:

$$hf0 = E2 - E1$$

This transition is used by atomic clock to build a time reference. This connotation forms the basis of atomic frequency standard which is used to generate accurate and precise time and frequency, enabling many communications, synchronization, and navigation systems in modern life [3].

---

[4]Atomic Oscillator, NIST.gov. Available online at https://www.nist.gov/pml/time-and-frequency-division/popular-links/time-frequency-z/time-and-frequency-z-am-b.

1.4.1 Atomic Frequency Standards and Atomic Clocks

The atomic frequency standard is a device which is part of a class of frequency standards characterized by extremely outstanding performance in terms of frequency stability and accuracy. The basic resonant system of these devices is an atom or a molecule experiencing a transition between two well-defined levels as discussed earlier. An extremely accurate atomic clock can be constructed by locking an electronic oscillator to the frequency of an atomic transition as discussed earlier. The frequencies associated with such transitions are so reproducible that the definition of the second is now tied to the frequency associated with a transition in cesium-133 (Cs-133):

$$1s = 9,192,631,770 \, \text{cycles of the standard Cs} - 133 \, \text{transition}$$

Prior to the definition of "second" as the measurement of the number of cycles of radiation from a particular cesium-133 transition in 1967, reference to second was based on astronomical observation. Cesium is one of the elements used in the atomic standard devices, other elements are hydrogen and rubidium. For these devices, the resonance in question is a transition between two energy levels inside the hyperfine structure of their ground state. In 2008, physicist at NIST (National Institute of Standard and Technology) developed a highly accurate atomic clock technology known as quantum-logic clock which provides better accuracy than existing atomic clocks.

The graph presented in Fig. 1.6 depicts oscillator comparisons in terms of accuracy and power [4].

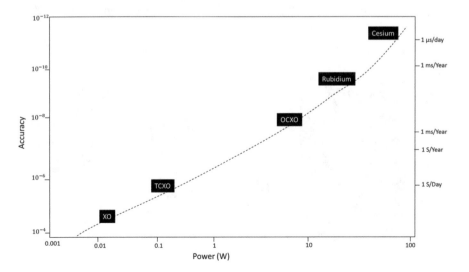

Fig. 1.6 Oscillator comparison graph

Since quantum logic clock is still under experimentation, its parameters are excluded from the comparison. The graph shows accuracy difference in several order of magnitude from the simple XO oscillator to cesium frequency standard. As the accuracy increases, so does the power requirement, size, and cost. Accuracy versus cost would be a similar relationship, ranging from about $1 for a simple XO to about $30,000 for a cesium standard.

1.4.1.1 Hydrogen Atomic Clock

Also known as hydrogen maser or hydrogen frequency standard, hydrogen atomic clock is an elaborate and expensive commercially available frequency standard or clock. The word *maser* stands for "**m**icrowave **a**mplification by **s**timulated **e**mission of **r**adiation." Masers operate at the resonance frequency of the hydrogen atom, which is 1,420,405,752 Hz [5].

In a hydrogen maser, hydrogen gas is pumped through a magnetic state selector that only allows atoms in certain energy states to pass through. Figure 1.7a depicts a sketch developed by "The European Space Agency" (esa) [6] and Fig. 1.7b represents inner works of active hydrogen maser device.

A hydrogen pump as shown in Fig. 1.7b pumps hydrogen gas through the magnetic selector. The atoms that pass through the magnetic selection gate enter a Teflon-coated storage bulb surrounded by a tuned, resonant cavity (a metal resonant structure). Once entered the bulb, some atoms drop to a lower energy level resulting in the release of photons of microwave frequency. These photons then stimulate other atoms to drop their energy level and in process release additional photons. The outcome is a self-sustaining microwave field that builds up in the bulb. Furthermore, the tuned cavity around the bulb helps to redirect photons back into the system to keep the oscillation going. The process produces a microwave signal that is locked

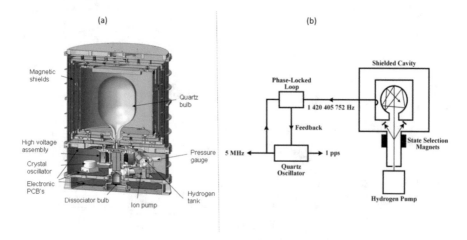

Fig. 1.7 The diagrammatical representation of hydrogen maser device [4–6]

to the resonance frequency of the hydrogen atom. This microwave signal is continually emitted as long as new atoms are fed into the system. A tunable quartz crystal oscillator is phased locked to this microwave signal through a downconversion received. The output frequency can be either 5 MHz or 1 MHz. It is to be noted that frequency stability of hydrogen maser decreases over time due to changes in the cavity's resonance frequency. Hence, a measurement of days or weeks may show hydrogen maser performance relatively below that of cesium atomic clock.

1.4.1.2   Rubidium Atomic Clock

The two most commonly used atomic clocks in recent years are rubidium and cesium. Both involve the locking of an electronic oscillator to the atomic transition. Rubidium is the lowest priced member of atomic clock which operates near 6.83 GHz of resonance frequency. Figure 1.8 illustrates a typical rubidium clock.

A microwave signal derived from the crystal oscillator is applied to the Rb1 vapor within a cell, forcing the atoms into a particular energy state. The optical beam from rubidium lamp is then pumped into the cell and absorbed by atoms forcing them into a separate energy state. As depicted in the figure above, the "photo cell" detector measures amount of the beam absorbed, and output is then used to tune a quartz oscillator to a frequency that maximizes the amount of light absorption

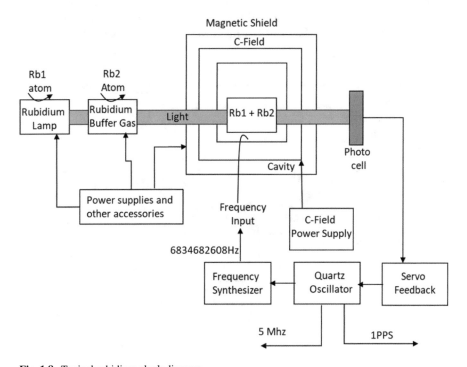

Fig. 1.8 Typical rubidium clock diagram

[7]. The quartz oscillator is then locked to the resonance frequency of rubidium and standard frequency output is derived from it. The rubidium oscillator continues to get smaller and less expensive. Their long-term stability is much better than crystal oscillator for price performance ratio.

1.4.1.3 Cesium Atomic Clock

The cesium oscillator is primary standard for atomic frequency and as discussed earlier, cesium isotopes transition is used to define SI second. The SI is a standard unit of measurement defined by the International System of Units (SI). We will discuss this further in the later part of this chapter. The resonance frequency of the cesium that determines the value of "second" is 9192, 631,770 Hz. An accurate cesium oscillator should be close to its nominal frequency without adjustment. There should not be any change in frequency due to aging.

Many of the commercially available cesium oscillators use cesium beam technology.

Figure 1.9 depicts cesium beam oscillator technology, inside vacuum cavity the cesium oven heats CS-133 atoms to gaseous state. Atoms from this gas leave the oven in a high-velocity beam that travels through a vacuum tube towards a pair of magnets known as state selection magnets. The pair magnets act as gate allowing only atoms of a particular magnetic state to pass through them. These atoms then entered to a microwave cavity where they are exposed to microwave frequency derived from a quartz oscillator. If the microwave frequency matches the cesium

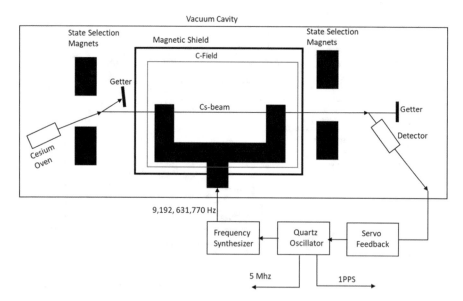

Fig. 1.9 Cesium beam oscillator diagram

resonance frequency, the atoms of cesium change their magnetic state. This cesium atom beam (Cs-beam) is then passed through another pair of magnets at the end of the tube. Those atoms that have changed their state while passing through microwave cavity will be allowed to reach the detector. Atoms that did not change their state are deflected from detector. The detector then produces a feedback signal that continually tunes the quartz oscillator. It is happened in way that maximizes atom state changes allowing greatest number of atoms to enter detector. Similar to rubidium oscillator, standard output frequencies are derived from the locked quartz oscillator.

1.4.1.4   Quantum Logic Clock

The quantum clock is a type of atomic clock where single ions are cooled with a laser and contained within an electromagnetic trap. It detects the energy state of a single aluminum ion and keeps time to within a second every 3.7 billion years. This new timekeeper will 1 day replace cesium frequency standard for internal standard (SI) timekeeping and SI unit of time "second" will be based upon transition of aluminum (AI) ions. Quantum clocks are more precise than atomic clocks. It is not affected by temperature or background noise from electric and magnetic fields.

To keep time, quantum-logic clock measures the vibration frequency of UV lasers. Unfortunately, even the best lasers can veer off their normal frequency by about one tick every hour. To keep the laser's timekeeping precise, its vibration must be anchored to something much more stable. That anchor is the vibration of an electrically charged aluminum atom, which vibrates at 1.1 Petahertz, or 1.1 quadrillion times a second.

The quantum clock was developed in 2010 by US National Institute of Standards and Technology (NIST). Till date quantum-logic clock has not been commercialized and mainly used for experimental purpose.

1.4.2   Miniaturization of Atomic Clock

Modern communication infrastructure and applications require precision timing that can be obtained through either of the two ways: GNSS based time that is ultimately powered by atomic clock and the direct use of atomic clock. For the former, resiliency cannot be guaranteed in case of war or environmental condition such as ionospheric disturbances, etc. On the other hand, miniaturized atomic clocks available today are cost effective. These clocks could be stationary to a given network helping anomalies to remain manageable. Over the past few decades, US Defense Advanced Research Projects Agency (DARPA) has invested heavily in the advancement and miniaturization of atomic clock technology, generating chip-scale atomic clocks (CSACs) that are now commercially available.

These miniaturized atomic clocks provide unprecedented timing stability for their size, weight, and power. However, the performance of these first-generation

CSACs are fundamentally limited due to the physics associated with their designs. Calibration requirements and frequency drift can generate timing errors, making it difficult to achieve the highest degrees of accuracy and reliability in a portable package [8]. Hence, DARPA is working towards an ultra-miniaturized low-power, atomic time and frequency reference unit that may provide 1000× improvement in key performance parameters over existing options (Fig. 1.10).

This ultra-miniaturized and ultra-low power atomic clocks will be useful for high-security Ultra High Frequency (UHF) communication and jam-resistant GPS receivers. The product can greatly improve the mobility and robustness of any military system or platform with sophisticated UHF communication and/or navigation requirement.

1.5 Unit of Time

According to time and frequency standards, there are three basic information: time of the day (TOD), time interval, and frequency. The "time of day" information consists of hours, minutes, and seconds, but also includes date (year, month, and day). The device that displays or records TOD is clocked clock. If a label is used by the clock to record the event of TOD, it is called time tag or timestamp. The timestamp can be used to synchronize the clocks in a communication infrastructure. Time interval, on the other hand, is duration or elapsed time between two events. Its unit measurement is "second." However, many critical communications infrastructures such as 5G mobile networks, financial networks, and smart grid

Fig. 1.10 An ultra-miniaturized atomic clock under development by DARPA (Courtesy: DARPA) [8]

require much shorter interval of time, such as milliseconds (1 ms = 10^{-3} S), micro-seconds (1 μs = 10^{-6} S), nanoseconds (1 ns = 10^{-9} S), and picoseconds (1 ps = 10^{-12} S).

Time is one of the seven base physical quantities defined by International System of Unit (SI). Others are *meter* for measurement of length, the *kilogram* for mass, the *ampere* for electric current, the *kelvin* for temperature, the *mole* for amount of substance, and the *candela* for luminous intensity. The unit of time defined by SI is "second" or often known as *SI second*. It was formally defined as 1/86,400 of the mean solar day based on the earth's rotational rate on its axis relative to Sun. That changed in 1967, when cesium second is defined. It was based on a measurement of the number of cycles of the radiation from a particular cesium-133 transition as discussed earlier. In 1967, the 13th General Conference of International Committee of Weights and Measures on Weights and Measures provisionally defined the second as 9192,631,770 cycles of cesium atomic clock's resonance frequency. The definition is stated as follows:

> The duration of 9,192,631,770 periods of the radiation corresponding to the transition between two hyperfine levels of the ground state of the cesium-133 atom [5].

The new definition meant that seconds are now measured by counting oscillations of electric fields that cause atoms to change state, and minutes and hours were now multiples of the second rather than divisions of the day. The benefits of this atomic timekeeping to our society are immense. Many critical infrastructures that we take for granted, such as global navigation satellite systems, mobile telephones, and the "smart grids" depend upon atomic clock accuracy.

The next information element "frequency" can be understood as the rate of repetitive event. For example, if "T" is the period of repetitive event, then frequency "f" is its reciprocal, $1/T$. Conversely, period is reciprocal to frequency, $1/f$. Hence, if period is time interval represented by second, then it is easy to see the relationship between time interval and frequency. The unit of frequency is hertz (Hz) defined as event or cycles per second. The larger multiples of Hz is Kilohertz (kHz), Megahertz (MHz), or Gigahertz (GHz) for which 1 kHz is one thousand (10^3) events per second, 1 MHz is one million (10^6) events per second, and 1 GHz is one billion (10^9) events per second.

Therefore, by counting seconds we can estimate date and TOD, and by counting events or cycles per second we can measure frequency.

1.6   Coordinated Universal Time (UTC)

The primary time standard by which the world regulates clocks and time is UTC. It is a 24-h time standard that uses highly precise atomic clocks combined with the Earth's rotation. Timing centers around the globe agreed to keep their time scales synchronized or coordinate and hence, the name coordinated universal time. It is the successor of Greenwich Mean Time (GMT). The UTC was defined by the International Radio Consultative Committee (CCIR), a predecessor organization of the ITU-TS, and is maintained by the Bureau International des Poids et Measures (BIPM).

Major metrology laboratories around the world routinely measure their time and frequency standard and send the data to BIPM at Sevres, France. The BIPM averages more than 200 atomic time and frequency standard dataset collected from more than 40 laboratories around the world including NIST. This averaging of data helps BIPM to generate two time scales: International Atomic Time or TAI and UTC. These time scales maintain SI second as closely as possible. TAI and UTC run at the same frequency; however, UTC differs from TAI by an integral number of seconds. This difference increases when leap second occurs.

The Earth's rotation, as measured by Universal Time 1 (UT1), is not as precise. The average length of time it takes for the planet to complete one rotation is 86,400.002 s while a day measured by atomic clocks is exactly 86,400 s. The discrepancy is due to Earth's rotation that slows because of a braking force applied by the gravitational pulls among the Earth, Sun, and Moon. In a year, this discrepancy of two milliseconds between UT1 and UTC grows to 1 s. Over a number of decades, this discrepancy could be 1 min and over thousands of years the clocks could be off by an hour. Hence, a leap second is added to UTC on June 30 or December 31st. The leap seconds have been added to UTC at a rate of slightly less than once per year, beginning in 1972.

The UTC is considered the ultimate standard for time of day, time interval, and frequency. Clocks that are synchronized to UTC provides same hour, minute, and second all over the world. Oscillators also syntonized to UTC to generate signals that serves as standard for time interval and frequency [5].

References

1. ECC. (2016). *Practical guidance for TDD networks synchronization* (ECC Report 2016).
2. ITU-T Rec G. 8260. (2020). *Definitions and terminology for synchronization in packet networks*. ITU-T.
3. Audoin, C., & Vanier, J. (1979). Atomic frequency standards and clocks. *Journal of Physics E: Scientific Instruments*. https://ui.adsabs.harvard.edu/abs/1976JPhE....9..697A/abstract
4. RCC. (2009). *Definition of frequency and timing terms, satellite navigation and timing system, and behavior and analyses of precision crystal and atomic frequency standards and their characteristics*. Document-214-09. Telecommunications and Timing Group, Range Commanders Council (RCC).
5. Lombardi, A. M. (2003). *Hydrogen maser, time and frequency section in encyclopedia of physical science and technology* (3rd ed.). Available online at https://www.sciencedirect.com/topics/earth-and-planetary-sciences/hydrogen-maser
6. esa. (2011). *Sketch of the space hydrogen maser*. The European Space Agency. Available online at http://www.esa.int/ESA_Multimedia/Images/2011/10/Sketch_of_the_Space_Hydrogen_maser
7. Lombardi, M. (2001). *Fundamentals of time and frequency in book: the mechatronics handbook*. Retrieved October 27, 2020, from https://www.researchgate.net/publication/269411932_Fundamentals_of_Time_and_Frequency
8. DARPA. (2019). *DARPA making progress on miniaturized atomic clocks for future PNT applications*. US Defense Advanced Research Projects Agency. Available online at https://www.darpa.mil/news-events/2019-08-20

Chapter 2
Timing Parameters

2.1 Introduction

In modern digital communications network, synchronization is often forgotten, taken for granted yet an essential element. For 5G, synchronization is essential to the performance of the network and its operational stability. To correctly design a network, synchronization consideration is de facto for most of critical infrastructure, e.g., 5G mobile networks financial networks, data center, and smart grid. Since much of telecom network is asynchronous due to increase in deployment of ethernet, network asymmetry should be an important consideration for such network design. For this purpose, understanding of various timing parameters, how these parameters apply to networks, timing source, and equipment are critical. Moreover, network design must adhere to ITU-T recommended time error budget for end-to-end traffic path.

This chapter explains need for synchronization and discusses various time error condition, budget and accuracy requirements. Understanding the time error and synchronization measurement technique presented in this chapter will help designer of network and equipment alike to build optimized device and network. The chapter explores industry standards such as applicable ITU-T recommendations for measuring time error in standalone setting as well as in network reference points. These fundamental concepts will help readership explore applicable synchronization techniques in more complex network scenarios of those that will be presented in proceeding chapters.

© The Author(s), under exclusive license to Springer Nature
Switzerland AG 2021
D. D. Chowdhury, *NextGen Network Synchronization*,
https://doi.org/10.1007/978-3-030-71179-5_2

2.2   The Need for Synchronization

Synchronization in digital communications networks (e.g., LTE or 5G mobile networks) is the process of aligning the time scales of network equipment and other transmission devices as such so that equipment operations occur at the correct time and correct order. To perform synchronization, the receiver clock at a network equipment or device must acquire and track the periodic timing information in a transmitted signal.

Figure 2.1 illustrates a typical LTE mobile network setup where eNodeB connects to cell tower radio unit and offload traffic over ethernet connection to packet network. The DCSG (Disaggregated Cell Site Gateway) is a network switching equipment that provides connectivity of eNodeB to packet network. The transmitted signal from eNodeB includes data that is clocked out at a rate determined by its transmitter clock. The output signals of eNodeB transition between zero and peak value containing clock information. Detecting these transitions allows clock to be recovered at the receiver, herein the clock recovery circuitry at DCSG. The received signal at the recovery circuitry of DCSG is used to write the received data into buffer with reduced jitter (we will discuss about jitter later in this chapter). Data is then read out of buffer for further processing and transported to digital bus within the DCSG. This process is identical in both directions: eNodeB to DCSG and DCSG to eNodeB.

In most cases, these equipment maintain the sense of time through a local clock and by counting the pulses of an internal crystal oscillator. However, there may be

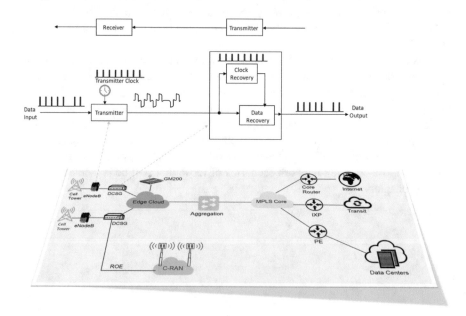

Fig. 2.1 A diagrammatical representation of transmitter and receiver clock

inherent inaccuracy in frequency (causing clock skew, we discussed it in Chap. 1) and phase of the crystal oscillator. The inaccuracy is influenced by the operating conditions and aging (resulting in clock drift as discussed in Chap. 1) [1]. As a result, these devices may deviate from a reference clock after a synchronization epoch. Such issues of poor synchronization manifest in a network as timing error.

2.3 Time Error

In our example at Fig. 2.1, we assumed that eNodeB has an accurate clock and the signal pulses out of the device are transmitted at precise intervals (pulse repetition period) which arrived at the receiver (herein DCSG, please refer to Fig. 2.1) with exactly the same time spacing. What if this is not the case and eNodeB in this example has an imperfect clock and the signals transmitted out of the device at a rate that is determined by the imperfect clock. In this case, signals arrive at DCSG slightly different times due to physical and electrical transmission processes. Assuming that DCSG also has imperfect recovery clock, this received signals which may be already corrupted by noise and phase delay distortion, detected and picked up by imperfect clock recovery circuits for processing, thus adding more noise and phase distortion to an already distorted signal. Such issue may result in *jitter* and *wander*. We will discuss jitter and wander later. However, there are several other time errors as well. For example, eNodeB in our example reported a time difference between its internal clock and the reference clock it received from network or through a built-in GNSS PRTC (Primary Reference Time Clock) timing module. Here time received through GNSS PRTC is configured standard time reference clock. The deviation of its local clock and the time indicated by the standard clock is the time error. Similarly, a network equipment clock accuracy can be measured using a test equipment that supplies reference clock and measures the differences. According to ITU-T recommendation G.810, time error can be mathematically defined as follows:

$$x(t) = T(t) - T_{ref}(t)$$

[where time error function $x(t)$ is the difference between a local clock generating time $T(t)$ and standard reference clock $T_{ref}(t)$].

The result of this time error can be either negative or positive. ITU-T recommendation G.8260 characterized time errors in four different categories:

- constant time error (cTE): It is the mean of the time error function and normally expressed as a single number, and compared to an accuracy specification, e.g., ±50 ns [2].
- dynamic time error (dTE): It is the random noise components of a referenced clock. It typically represents the dynamic nature of clocks in timing distribution systems containing the sum of all its unpredictable components such as GNSS timing error, time-stamping errors, queues/buffers/memories, PDV, traffic pat-

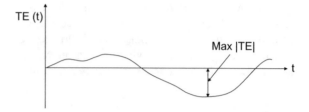

Fig. 2.2 A diagrammatical representation of Max |TE| [4]

terns, noise, oscillators' frequency variations and temperature dependencies, among other phase noise sources. dTE(t) is described as the variable part [3].

- maximum absolute time error (Max |TE|): ITU-T G.8260 defines this type of time error as maximum absolute value of the time error function of a synchronized clock. It is typically expressed as a single number and compared to an accuracy specification, e.g., ±100 ns (Fig. 2.2).

2.3.1   Jitter and Wander

A jitter can be understood as deviation from periodicity of assumed periodic signal and often in relation to a reference clock signal. A timing jitter can be understood as deviation of clock signal edge from its ideal location. ITU-T recommendation G.810 defines timing jitter and wander as follows:

Jitter: "The short-term variations of the significant instants of a timing signal from their ideal positions in time (where short-term implies that these variations are of frequency greater than or equal to 10 Hz)."

Wander: "The long-term variations of the significant instants of a digital signal from their ideal position in time (where long-term implies that these variations are of frequency less than 10 Hz)."

According to these definitions of jitter and wander, jitter can be understood as phase variations that occur at a rate greater than 10 Hz while wander occurs at a rate less than 10 Hz. Following is a diagrammatical representation of timing jitter and wander as presented in ITU-T G.8260 that depicts deviation of edges of the timing signal or phase variation causing jitter and wander (Fig. 2.3).

Jitter can be measured and expressed in multiple ways, here is a list of different types of jitters:

- Period Jitter: It is the deviation in cycle time of a clock signal with respect to the ideal period over a number of randomly selected cycles. The period jitter measures the maximum deviation of clock period of a clock cycle in the waveform over 10,000 clock cycles.
- Cycle to Cycle (C2C) Jitter: The C2C jitter is defined as the maximum difference between any two adjacent clock periods [6]. It is defined in JEDEC Standard 65B as the variation in cycle time of a signal between adjacent cycles, over a random sample of adjacent cycle pairs [7]. Figure 2.4 depicts deviation of clock's output

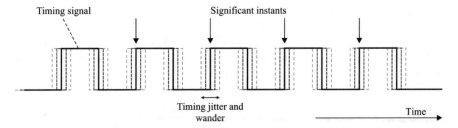

Fig. 2.3 A diagrammatical representation of timing jitter and wander (courtesy: ITU-T rec. G.8260) [5]

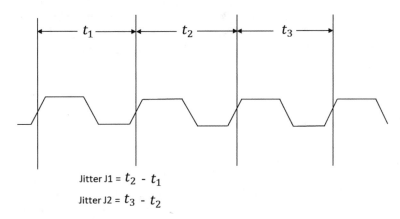

Jitter J1 = t_2 - t_1

Jitter J2 = t_3 - t_2

Fig. 2.4 A diagrammatical representation of C2C jitter [6]

transition from its previous position. Here jitter J1 and J2 are presented to reflect the deviation and values measured over multiple cycles.

- Long-Term Jitter: It measures deviations in clock's output from ideal position over a period of time. Actual duration depends on application and the clock frequency. In a typical personal computer motherboard, this duration is 10–20 μs.
- Phase Jitter (Phase Noise): It is usually described as a continuous noise plot over a range of frequencies or a set of noise values at different frequency offsets. This type of jitter is measured as the integration of phase noises over a certain spectrum and expressed in seconds.

2.3.2   Packet Delay Variation (PDV)

ITUT-T G.8260 defines this type of phase variation as "significant instance" that is equally applicable to packet timing signal. For a packet network, this packet timing signal is encoded in the series of time-critical packets known as "event" packets. During transmission in a packet network, event packets still contain significant instances of the timing signal which is subject to variation due to network condition. This variation is known as Packet Delay Variation or PDV. The PDV can be caused by several factors including transmission delay, processing delay, and buffering delay. Transmission delay occurs due to velocity of signal between two endpoints in a network and distance between them. Such delay is generally induced by underlying medium technologies, e.g., wireless or wired medium. The processing delay is induced by processing of timing packet by network equipment. Similarly, as packet traverses through various switches and routers in a given network, these routers and switches queue the packets for processing, thus inducing buffering delay. These PDV components have different PDV ranges. The transmission delay could induce a PDV of sub-microseconds while processing delay by CPU may induce PDV in the 1–10 µs range. However, buffering delay could induce PDV in a wide range typically 10 µs to 1 ms. PDV becomes a serious concern for packet network where timing protocol such as PTP (Precision Time Protocol) or IEEE1588 is used to distribute timing information.

2.4   Synchronization Measurement

Considering various time error that may occur in a digital communications network, time delay synchronization measurement is imperative. Inadequate or poor synchronization compromises quality of service, leading to such impairments as data retransmission, digital video freeze and distortion, and severe degradation of encrypted services. For mobile transport, such impairment may cause significant network issues and operational failure in the network. Such synchronization time delay can be measured in two ways: first measurement is related to the frequency and concerns the evaluation of the tolerance of the system to the frequency offset. In this process, behavior of the network element is observed. The second is related to phase and concerned with measures of jitter and wander.

2.4.1   Phase Measurement

As stated in Chap. 1, phase is associated with a repetitive waveform. In such signal (as depicted in Fig. 2.5), waveshape repeats itself once the period of repetition elapses.

For a given time, phase is the fractional portion of the period that completed. It is commonly expressed in degrees or radians. The full cycle is 360° or 2 π. Hence, phase is zero when cycle is beginning and phase is 180° when cycle is half complete. Please note, phase is defined as the portion of the cycle that is complete and depends on where the beginning of the cycle taken to be. A deviation in phases of received signals could indicate severe issues in synchronization of a network. As discussed in Chap. 1, phase alignment is a key criterion of phase and time

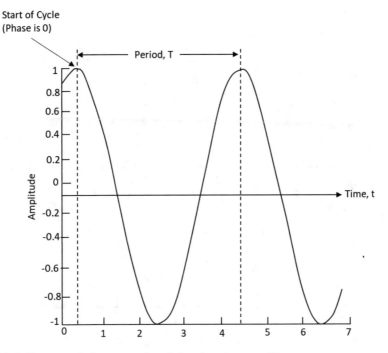

Fig. 2.5 A diagrammatical representation of phase in a sine wave [8]

synchronization for a communications network. In addition, phase is an important measurement in many applications, for example, in rotating machinery phase measurement is used to detect fault (Fig. 2.6).

The phase measure consists in the comparison of the electrical signal of the clocks under examination with the reference one. A phase detector is typically used to measure the phase. It senses when a signal passes through a set voltage threshold, timing the threshold crossing relative to a reference. The difference in phase between the two signals (known as phase shift) is a result of the electrical characteristics of the device under test. The reference clock is obtained from the same clock under test by filtering the frequency above the threshold of frequency equal to 10 Hz.

2.4.1.1 Time Interval Error (TIE)

The TIE is defined as phase deviation between the signal being measured and the reference clock. It is measured in nanosecond (ns). ITU-T recommendation G.810 defines TIE as "the difference between the measure of a time interval as provided by a clock and the measure of that same time interval as provided by a reference clock." The deviation can be presented mathematically as the time interval error function TIE $(t; \tau)$, which can be expressed as [9]:

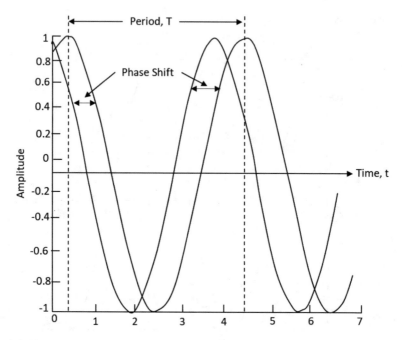

Fig. 2.6 Phase measurement: deviation of two relative signals [8]

$$\text{TIE}(t;\tau) = \left[T(t+\tau) - T(t)\right] - \left[T_{\text{ref}}(t+\tau) - T_{\text{ref}}(t)\right] = x(t+\tau) - x(t)$$

[where τ is the time interval, usually called observation interval].

2.4.1.2   Minimum Time Interval Error (MTIE)

It is historically been one of the main time-domain quantities considered for the specification of clock stability requirements in telecommunications standards. The MTIE is a rough measure of the peak time deviation of a clock with respect to a known reference.

2.4.1.3   Time Deviation (TDEV)

This is a measure of the expected time variation of a signal as a function of integration time. TDEV can also provide information about the spectral content of the phase (or time) noise of a signal [9]. While MTIE is a kind of peak detector, TDEV by contrast is a "rms" measure of wander over various integration times. It was specifically designed for the characterization of wander noise processes operating within the network. Both MTIE and TDEV characterize wander over a range of values ranging from short-term wander to long-term wander [10].

2.5   Measuring Time Error in Packet Network

Now that we understand time error (TE) and associated parameters for characterization of time error and measurements, let's explore how these parameters are applied to a typical telecom packet networks. It is to be noted that modern telecom network including mobile network frontend and the core networks are based on ethernet and hence, reference to packet network considers ethernet as underlying communication medium. There may be some variation, e.g., DWDM (Dense Wave Division Multiplexing) and PON (Passive Optical Network) for aggregation of mobile backhauls and connectivity to the core. However, irrespective of technological differences at physical medium level, these technologies have one thing common they carry packet for transport.

ITU-T recommendation G.8271.1 has specified TE (time error) limit for such packet networks at given network reference point (please refer to Fig. 2.7).

Reference point A shows input of PRTC (Primary Reference Time Clock) to T-GM (Telecom Grand master clock). A Grand master clock or T-GM is mostly associated with precision time protocol (PTP) based packet network. A T-GM provides reference clock for a packet network. In real-world deployment, all T-GM includes PRTC input built-in. Such PRTC includes GNSS (Global Navigation Satellite System) based clock input with wired clock output such as 1PPS and 10 MHz along with SyncE and PTP support. We will discuss details about T-GM later in the proceeding chapters. For now, T-GM should be understood as main clock reference input for a given packet network. The GM200 is a T-GM product of Trimble, such products are very cost-effective and have good performance in terms of reference clock output for a small macro-aggregation. The ITU-T G.8271.1 network reference points are overlaid on a practical LTE/LTE-A mobile network. Maximum hop count for T-BC (Telecom Boundary Clock) is 5 between reference point B and C. The function of T-BC in this scenario is to act as master clock for endpoint devices that implement slave clock (T-TSC). Similar to the T-GM, T-BC may implement PRTC inputs, but T-BC is relatively less accurate than T-GM and

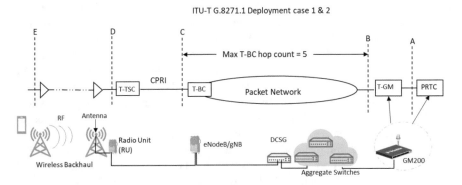

Fig. 2.7   ITU-T G.8271.1 network reference point and time error measurement [11]

uses T-GM reference clock to adjust its clock in case of clock discrepancies. Use of PRTC input is not mandatory for T-BC. In addition, T-BC also provides PTP with backward compatibility for NTP (Network Time Protocol).

Figure 2.7 merges deployment case 1 and 2 identified in ITU-T G.8271.1 and overlaid on top of real-world deployment. Please note that reference point D and E can be moved further to the right depending upon whether eNodeB includes a T-BC, transparent or a slave clock embedded in it. Latest eNodeB includes embedded timing modules that may include an embedded T-BC with PRTC capabilities. In that case, reference point E can be moved to wireless backhaul scenarios. However, reference points "A to D" are important for consideration of network design. ITU-T G.8271.1 defines following time error network limit for these reference points.

2.5.1 Time Error Limit at Network Reference Point A

In cases where T-GM has implemented PRTC, the combined solution should be within 100 ns or better for PRTC-A verified against primary time standard (e.g., UTC). ITU-T G.8272 defines two types of PRTC: PRTC-A and PRTC-B. PRTC-B has better accuracy than PRTC-A.

There are two types of errors induced by noise that are applicable to PRTCs:

- constant time error or time offset at the output of PRTC compared to the applicable primary time standard (e.g., UTC),
- phase error (wander and jitter) produced at its output.

Phase error is generally measured through the calculation of MTIE and TDEV. For the combined PRTC-A/PRTC-B and T-GM function, time error samples are measured through a moving-average low-pass filter of at least 100 consecutive time error samples. This filter is applied by the test equipment to remove errors caused by timestamp quantization, or any quantization of packet position in the test equipment, before calculating the maximum time error [12]. The accuracy of PRTC-B should be within 40 ns or better. The maximum absolute time error for PRTC-A is max|TE| ≤ 100 ns. ITU-T G.8272 defines wander measurements for PRTC-A and PRTC-B. The tables below depict MTIE and TDEV measurements for PRTC-A and PRTC-B as ITU-T G.8272 (Tables 2.1 and 2.2).

It is to be noted that where a T-GM (with embedded PRTC) has 1PPS interface, MTIE is applicable to 1 s observation period. The ideal setup for the measurement of MTIE and TDEV presented herein should have PRTC in locked mode as described in ITU-T G.810 and G.8272. The locked mode is the operational condition in which PRTC (herein Trimble's GM200 T-GM as shown in Fig. 2.8) is fully locked to incoming reference signal and not in warmup condition.

The same condition of locked mode and setup as described above is applicable to wander expressed in TDEV measurement. ITU-T recommendation G.8272 defines TDEV limit as depicted in the tables below (Tables 2.3 and 2.4).

Table 2.1 Wander generation (MTIE) for PRTC-A

MTIE limit (µs)	Observation Interval τ (s)
$0.275 \times 10^{-3} \tau + 0.025$	$0.1 < \tau \leq 273$
0.10	$\tau > 273$

Table 2.2 Wander generation (MTIE) for PRTC-B

MTIE limit (µs)	Observation Interval τ (s)
$0.275 \times 10^{-3} \tau + 0.025$	$0.1 < \tau \leq 54.5$
0.04	$\tau > 54.5$

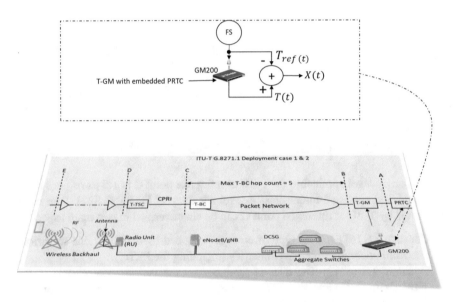

Fig. 2.8 A diagrammatical representation of wander measurement at PRTC in locked mode (courtesy: ITU-T G.810, Fig. 1a)

For intrinsic jitter measurement at PRTC, 10 MHz output interface at T-GM should be measured at 60-s interval and shall not exceed 0.01 UIpp of unit interval (UI) specification [13].

2.5.2 Time Error Limit at Network Reference Point B

For the T-GM with embedded PRTC, TE parameters presented for reference point A are applicable to reference point B. ITU-T recommendation G.8271.1 did not specify TE parameters for T-GM that does not integrate PRTC and PRTC is external to the device.

Table 2.3 Wander generation (TDEV) for PRTC-A

TDEV limit (ns)	Observation interval τ (s)
3	$0.1 < \tau \le 100$
$0.03\,\tau$	$100 < \tau \le 1000$
30	$1000 < \tau \le 10000$

Table 2.4 Wander generation (TDEV) for PRTC-B

TDEV limit (ns)	Observation Interval τ (s)
1	$0.1 < \tau \le 100$
$0.01\,\tau$	$100 < \tau \le 500$
5	$500 < \tau \le 100000$

Table 2.5 dTE network limit expressed in MTIE

MTIE limit (μs)	Observation Interval τ (s)
$100 + 75\,\tau$	$1.3 < \tau \le 2.4$
$277 + 1.1\ \tau$	$2.4 < \tau \le 275$
580	$275 < \tau \le 10000$

2.5.3 Time Error Limit at Network Reference Point C

The accuracy level applicable for T-BC at reference point C is 1.5 μs as per ITU-T G.8271.1 and specified under class 4 level accuracy at ITU-T G.8271. The noise generated by T-BC at reference point C can be characterized in two main aspects:

- Constant TE (cTE): This type of error is produced by the chain due to various fixed and uncompensated asymmetries (including the PRTC).
- Dynamic TE (dTE): The dTE is produced by the various components of the T-BC chain (including the PRTC)—this noise can be classified as low or high frequency noise, with components below or above 0.1 Hz, respectively [11].

The maximum absolute TE network limit is max|TE| $\le$ 1100 ns. The network limit for dynamic low frequency TE, dTE L (t), is specified in ITU-T G.8271.1 in terms of MTIE as depicted in Table 2.5.

Please note, specification for TDEV at reference point C is not provided in ITU-T G.8271.1 and subject to further study.

2.5.4 Time Error Limit at Network Reference Point D

The TE limit at reference point D is similar to reference point C in most cases. There is no TE limit specified in ITU-T G.8271.1 for network reference point E. Please note reference point E may vary depending upon deployment scenario as discussed earlier.

2.5.5   Phase Wander Measurement at PRTC

ITU-T G.8272 recommends the techniques of measuring phase wander of a PRTC relative to atomic clock frequency standard referred to as PRC (Primary Reference Clock), e.g., Cesium clock. A time interval counter can be used to compare the phase of a 1PPS output signal from the PRTC to that of a PRC as depicted in figure below. In this experimentation, T-GM with embedded PRTC is considered, e.g., Trimble's GM 200 (Fig. 2.9).

It is understood that phase wander in a cesium atomic clock is extremely low although it may have slight offset to UTC frequency. For a PRC reference, this offset should be 1 part in 10^{11} . However, this offset must be removed to reveal the wander performance of T-GM based PRTC. ITU-T G.8272 suggests using two PRCs in a three-way comparison as shown in Fig. 2.9 to distinguish between wander of PRTC and that of PRC.

2.6   Time Error Budget

Traditionally, network environment was tolerant to millisecond level delay for example, billing and alarm for a telecom network only required hundreds of milliseconds timing accuracy and much of the earlier deployment relied on wired timing distribution requiring mainly frequency synchronization. That requirement has changed overtime as telecom environment started to use both FDD and TDD spectrums requiring both phase and time synchronization as well as precision timing accuracy. ITUT-T G.8271.1 which mainly defines TE for LTE environment categorized TE requirement for telecom network. Such categorization emphasizes on fronthaul and backhaul mobile network and suggests the corresponding time and phase requirements for various classes of network endpoints. Figure 2.10 depicts a mobile edge network and corresponding classes of devices.

Overall network as ITU-T should have a maximum of 1.5 µs (1.1 µs preferred) for end-to-end mobile transport. The estimate is considered with count from direct PRTC input from T-GM (Telecom Grand master). The node 1–5 represent synchronization supply chain suggesting up to 5 T-BC hop count but in reality some of the US tier1 provider has deployed only two T-BC hop count. This node (1–5) hop count applies to different base stations network, whereas node 6a/b/c, 7a/b/c, 8a/b/c, and 9 represent synchronization supply chain that is applicable to specific base station only. The best way to understand this concept is that, if node 5 considered a base station than remaining nodes from 6 to 9 are part of its fronthaul cluster. For example, node 9 and 8b could be radio units which are up to 9 km apart ultimately connecting to same base station and their TAE budget is <260 ns. Another important consideration for network design is to consider classes of accuracy and their corresponding applications are shown in the tables below. This understanding is very

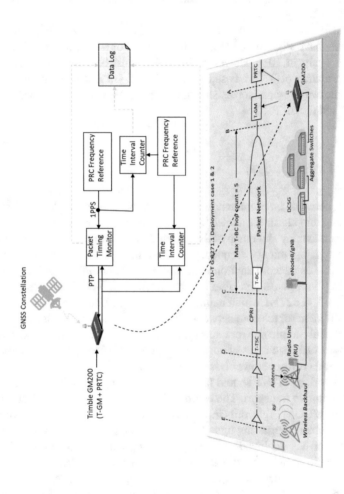

Fig. 2.9 Typical setup of measuring phase wander of a PRTC [9]

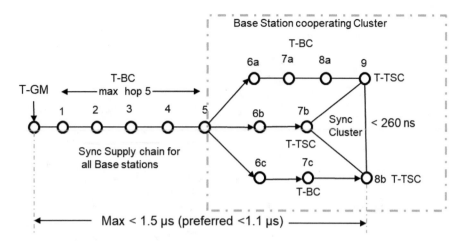

Fig. 2.10 A diagrammatical representation of time error budget for a typical telecom mobile network environment [11]

Table 2.6 Time and phase requirement as per ITU-T G.8271

Class level of accuracy	TE requirements	Typical applications
1	500 ms	Billing and alarm
2	100–500 μs	IP delay monitoring. Synchronization signal block (SSB)-measurement timing configuration (SMTC) window
3	5 μs	LTE-TDD (large cell). Synchronous dual connectivity (for up to 7 km propagation difference between eNBs/gNBs in FR1)
4	1.5 μs	UTRA-TDD, LTE-TDD (small cell), NR TDD, WiMAX-TDD (some configurations). Synchronous dual connectivity (for up to 9 km propagation difference between eNBs/gNBs in FR1) (note 2). New radio (NR) intra-band noncontiguous and inter-band carrier aggregation, with or without multiple input multiple output (MIMO) or transmit (TX) diversity
5	1 μs	WiMAX-TDD (some configurations)
6	X ns	Various applications, including location-based services and some coordination features. Values of x vary as ITU-T G.8271 appendix II. An example is, "NR MIMO or TX diversity transmissions, at each carrier frequency" that needs 65 ns accuracy

useful when designing a network although much of the TE requirements are specific to telecom environment (Table 2.6).

Table 2.7 Time and phase requirement for cluster-based synchronization

Class level of accuracy	Maximum relative time error requirement	Typical applications
3A	5 µs	LTE MBSFN
4A	3 µs	NR intra-band noncontiguous (FR1 only) and inter-band carrier aggregation; with or without MIMO or TX diversity
6A	260 ns	LTE intra-band noncontiguous carrier aggregation with or without MIMO or TX diversity, and inter-band carrier aggregation with or without MIMO or TX diversity. NR intra-band contiguous (FR1 only) and intra-band noncontiguous (FR2 only) carrier aggregation, with or without MIMO or TX diversity
6B	130 ns	LTE intra-band contiguous carrier aggregation, with or without MIMO or TX diversity. NR (FR2) intra-band contiguous carrier aggregation, with or without MIMO or TX diversity
6C	65 ns	LTE and NR MIMO or TX diversity transmissions, at each carrier frequency

Please note Table 2.7 is only applicable to fronthaul mobile network scenarios, some classes of accuracy could be internal to a device, e.g., 6C which is applicable to a NR MIMO antenna device.

References

1. Mahmood, A., Gidlund, M., & Ashraf, I. M. (2018). *Over-the-air time synchronization for URLLC: requirements, challenges and possible enablers.* ResearchGate.
2. Calnex. (2016). *What is time error? Time error in brief.* Calnex Solutions Ltd.
3. Polo, M. I. (2016). *Can constant time error (cTE) be "measured"? A practical approach to understanding TE = cTE + dTE.* VeEX, Inc. Available online at https://download.veexinc.com/TX320s/Technical-Notes/6688/Measuring_Constant_Time_Error-cTE_D08-00-030_A00.pdf
4. Ruffini, S. (2013). *Q13/15 AR. time sync network limits: status, challenges.* Joint IEEE-SA and ITU Workshop on Ethernet.
5. G.8260. (2020). *Definitions and terminology for synchronization in packet networks.* ITU-T G.8260. International Telecommunication Union.
6. Application Note. (2010). *Understanding SYSCLK jitter.* Freescale Semiconductor. Available online at https://www.nxp.com/docs/en/application-note/AN4056.pdf
7. SiTime. (n.d.). *Clock jitter definitions and measurement methods.* SiTime.
8. O'Shea, P. (2000). *Phase measurement.* CRC Press, LLC.
9. G.810. (1996). *Definitions and terminology for synchronization of networks.* ITU-T Recommendation G.810. International Telecommunication Union.
10. Symmetricom. (2003). *Measuring and monitoring synchronization in a network.* Symmetricom.
11. G.8271.1. (2020). *Network limits for time synchronization in packet networks with full timing support from the network.* ITU-T G.8271.1/Y.1366.1 (03/2020). International Telecommunication Union.
12. G.8272. (2018). *Timing characteristics of primary reference time clocks.* ITU-T G.8272/Y.1367 (11/2018). International Telecommunication Union.
13. G.811. (2016). *Timing characteristics of primary reference clocks: Amendment 1.* ITU-T G.811 (04/2016). International Telecommunication Union.

Chapter 3
The Primary Reference Source (PRS)

3.1 Introduction

Central to time synchronization is the primary clock source known as PRS (Primary Reference Source). The PRS standard was developed by American National Standard Institute (ANSI) in 1987. The ANSI standard provided a guideline for primary reference clock accuracy and hierarchy level deployments of clocks and nodes in a network. The hierarchy of clock distribution defined in ANSI is known as "stratum." From historical perspective, the stratum level of hierarchy was used for effective network time distribution mechanism in TDM networks such as SONET/SDH network. However, it was also implemented in packet network timing distribution for NTP (Network Time Protocol) based network. The stratum model of hierarchical clock distribution is still valid and applicable, but it remains at the discretion of network professional for effective synchrony design. The concept of PRS evolved into ITU-T Standards which defined PRC (Primary Reference Clock) for SONET/SDH environment and PRTC (Primary Reference Time Clock) for modern packet networks (e.g., Ethernet) that uses PTP (Precision Time Protocol) as precision time distribution mechanism.

The networks need dependable and traceable primary reference time standard that was traditionally available through atomic frequency standard. However, atomic clock may not be viable in cost-effective distributed synchrony design and thus, GNSS-based PRTC integrated with PTP and/or NTP has become a common tool for synchronicity. In most of the modern networks, a combination of atomic frequency standard (delivered by central atomic clock) and GPS/GNSS-based primary reference time standards are used. The latter is obtained through PRTC apparatus. In this chapter, various PRS apparatuses are discussed along with network design technique offered for each. Since that PRTC is becoming mainstream for PRS in most network design, this chapter also explores PRTC in detail with various satellites constellations that provides radio signals to enable PRTC to obtain traceable

D. D. Chowdhury, *NextGen Network Synchronization*, https://doi.org/10.1007/978-3-030-71179-5_3

primary reference clock. The concept of PRC, PRTC, and satellite constellations will enable readership to better design product and networks for high-precision time distribution, thus helping them to build optimized network for the future.

3.2   Primary Reference Source (PRS)

All clock in a synchronized digital communications network must be referenced, or traceable, to a Primary Reference Source (PRS). The PRS is a master clock for a network that is capable to maintain a frequency accuracy of better than $1 \times 10\text{--}11$ [1] meaning only one error can occur in 10^{11} parts [2]. American National Standard Institute (ANSI) defined PRS and its hierarchy level in ANSI T1.101-1987 standard. The specification defines four stratum levels and minimum performance requirement for network synchronization. One class of PRS is a stratum 1 clock. A stratum 1 clock, by definition, is a free running clock. It does not use a timing reference to derive or steer its timing. Stratum 1 clocks usually consist of an ensemble of cesium atomic standards.

However, a PRS does not need to be implemented with primary atomic standards but frequency accuracy level discussed must be met. Today, GPS (Global Positioning System) clocks are used as PRS in most cases. The GPS and GNSS (Global Navigation Satellite System) may have onboard cesium clock or they synchronize with ground station that provides atomic frequency standard. Network clock devices such as T-GM (Telecom Grand Master) that synchronizes with GPS/GNSS for UTC time may also have quartz oscillators that are steered by timing information obtained from GPS/GNSS constellations. They are not considered stratum 1 since they are steered but are classified as PRSs. These clocks are able to maintain an accuracy within a few parts in 10--13 to a few parts in 10--12 [2].

There are two other terms for PRS, PRC (Primary Reference Clock) defined by ITU-T recommendation G.811 and PRTC (Primary Reference Time Clock) defined by ITU-T Rec. G.8272/Y.1367. The PRTC is mainly taking GPS/GNSS UTC input but in case it loses frequency and time references, it should either rely on its oscillator or get an external input from PRC. In this case, a PRC can be an atomic frequency reference standard (standalone cesium or rubidium clock). Depending upon applications, a standalone cesium or rubidium can be set up for time resiliency in a given communications infrastructure.

3.2.1   The PRS Timing Distribution Hierarchy

While discussing PRS, I referred various stratum clock levels for time distribution. ANSI T1.101-1987 standard defined four level of hierarchies for stratum clock. It was originally defined for the North American Telephone System as a hierarchy of clocks model to distribute time synchronization across the telecommunications

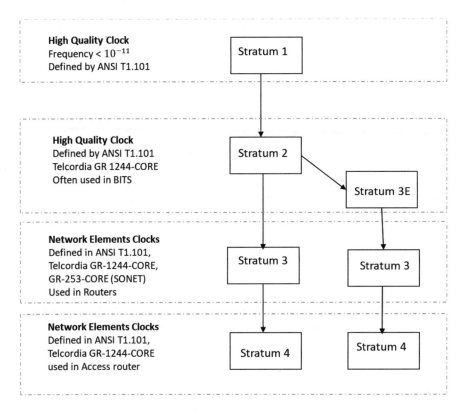

Fig. 3.1   PRS Clock hierarchy defined in ANSI T1.101-1987

network. The ANSI T1.101 is not in force today yet stratum levels are closely followed by other related standards (e.g., ITU-T) in terms of frequency accuracy
requirements. The hierarchy level of stratum clocks is illustrated in Fig. 3.1.

3.2.1.1   Stratum 1

As discussed, Stratum 1 is a completely autonomous source of timing with no other
input and acts as PRS for the network infrastructure. It can be standalone atomic
frequency standard (clock) or GPS/GNSS-based UTC reference which in turn uses
atomic clock onboard of the satellite or at ground station for clock reference. A
PRTC device can either use a reference oscillator (OCXO), GNSS UTC or both for
resilient time. An atomic standard like Cesium Beam or Hydrogen Maser may also
be the preferred choice. However, having a standalone atomic frequency reference
is very costly and in many situations distributed GPS/GNSS-based PRTC is used as
PRS for time distribution in a digital communications network. A Stratum 1 clock
may control strata or clock levels of 2, 3E, 3, 4E, or 4 clocks. A Stratum 2 clock may
drive strata 2, 3E, 3, 4E, or 4 clocks. A Stratum 3E clock may drive strata 3E, 3, 4E,

or 4 clocks. A Stratum 3 clock may drive strata 3, 4E, or 4 clocks. A Stratum 4E or 4 clock is not recommended as a source of timing for any other clock system [1].

3.2.1.2 Stratum 2

The main purpose of the Stratum 2 system is to track an input under normal operating conditions. It maintains the last best estimate of the input reference frequency during impaired operating conditions and works in tandem with other stratum 2 clocks in the network. An example of stratum 2 clock is rubidium clock or OCXO oscillator. The Stratum 2 clock may drive strata 2, 3E, 3, 4E, or 4 clocks but no other way around. Network synchronization design must follow this rule.

Table 3.1 shows Stratum 2 requirements.

3.2.1.3 Stratum 3

Stratum 3 clock system tracks an input as in Stratum 2, but over a wider range. It requires a minimum adjustment (tracking) range of $\pm 4.6 \times 10^{-6}$. The short-term drift is less than 3.7×10^{-7} in 24 h. This is about 255 frame slips in 24 h while the clock is holding. A lower cost single oven OCXOs and non-oven based temperature compensated crystal oscillators (TCXOs) can be employed in stratum 3 and stratum 4 based clocks. Table 3.2 depicts stratum 3 clock requirements.

3.2.1.4 Stratum 3E

Stratum 3E is defined by Bellcore documents as an upgrade clock standard of stratum 3 that takes into account SONET equipment requirements. It tracks input signals within 7.1 Hz of 1.544 MHz from a Stratum 3 or better source.

The drift with no input reference is less than $1 \times 10{-8}$ in 24 h. This is much different than Stratum 3. Stratum 3 is about 255 frame slips in 24 h, while Stratum 3E is only four slips. The following table depicts stratum 3E requirements (Table 3.3).

According to Telcordia GR-1244 standard, stratum 3E requirements on filtering of wander and holdover are significantly tighter than the stratum 3 requirements. The specification recommends that stratum 3E clocks be the minimum clocks used in BITS applications. In addition, it is recommended that stratum 3E or higher quality clocks not be used in any network element other than a BITS (e.g., it is recommended that transport NEs use stratum 3 or lower quality clocks).

Table 3.1 Stratum 2 Clock requirements

Level	Free-Run accuracy	Holdover Stability	Minimum pullin and hold-in range	Time to first frame slip*
Stratum 2	$\pm 1.6 \times 10^{-8}$	$\pm 1 \times 10^{-10}$ /day	$\pm 1.6 \times 10^{-8}$	7 days

* A slip is a measure of synchronization error, it is a frame (193 bits) shift in the time difference between two signals.

Table 3.2   Stratum 3 clock requirements

Level	Free-Run accuracy	Holdover stability	Minimum pullin and hold-in range	Time to first frame slip*
Stratum 3	±4.6 × 10⁻⁶	±3.7 × 10⁻⁷/day	±4.6 × 10⁻⁶	6 min (255 in 24 h)

Table 3.3   Stratum 3E Requirements

Level	Free-Run accuracy	Holdover Stability	Minimum pullin and hold-in range	Time to first frame slip*
Stratum 3E	±1.0 × 10⁻⁶	±1 × 10⁻⁸/day	±4.6 × 10⁻⁶	3.5 h

Table 3.4   Stratum 4E requirements

Level	Free-Run accuracy	Holdover stability	Minimum pullin and hold-in range	Time to first frame slip*
Stratum 4E	±32 × 10⁻⁶	Same as accuracy	±32 × 10⁻⁶	Not yet specified
Stratum 4	±32 × 10⁻⁶	Same as accuracy	±32×10–6	N/A

Note (*): To calculate slip rate from drift, one assumes a frequency offset equal to the above drift in 24 h, which accumulates bit slips until 193 bits have been accumulated. Drift rates for various atomic and crystal oscillators are well known, and are not usually linear or not necessarily continually increasing [1]

3.2.1.5   Stratum 4

Stratum 4 clock tracks an input as in Stratum 2 or 3, except that the adjustment and drift range is 3.2×10^{-5}. Also, it has no holdover capability and, in the absence of a reference, free runs within the adjustment range limits. The time between frame slips can be as little as 4 s.

Stratum 4E is proposed as a new customer premises clock standard that allows a holdover characteristic that is not free running. This new level, intended for use by customer provided equipment in extending their networks, is not yet standardized (Table 3.4).

3.2.2   Primary Reference Clock (PRC)

Prior to the industry wide acceptance of Ethernet, SONET and SDH were dominant technology physical medium technology. Unlike Ethernet both SONET and SDH are Layer 1 technology (in terms of OSI reference model), whereas ethernet combines both layer 1 and layer 2 (in terms of OSI reference model) in its layered architecture. SONET and its European standard technology "SDH" are a synchronous technology and require frequency synchronization to keep endpoint synchronize. Although we rarely encounter SONET and SDH technology for optical transport [which mostly replaced by DWDM (Dense Wave Division Multiplexing) and PON (Passive Optical Networking)], remnant of these technologies still exist in many network setups. ITU-T G.811 defined the PRC (Primary Reference Clock) and ITU-T G.803 defined the hierarchy of the clock mainly for the purpose of SONET/ SDH network synchronization. However, PRC as defined in ITU-T and its hierarchy is still in use in many network scenarios. Specifically, PRC as PRS is still valid and equally important as its counterpart PRTC (Primary Reference Time Clock).

ITU-T G.811 defines PRC as "a typical PRC provides the reference signal for the timing or synchronization of other clocks within a network or section of a network." The PRC also provides reference signal to slave clock defined in ITU-T G.812. An example of such slave or node clock is Stand-Alone Synchronization Equipment (SASE). However, slave for the purpose of this definition could be part of SDH cross-connect. The PRC should be maintained long-term accuracy at 1 part in 10^{11} or better with verification to coordinated universal time (UTC). A PRC could be a cesium or rubidium clock or high-quality OCXO as long as the long-term and short-term accuracy those specified in G.811 can be met by these oscillators.

3.2.2.1   PRC Hierarchy

The ITU-T Standards collectively provides PRC and node clock specifications that forms the PRC clock hierarchy as illustrated in Fig. 3.2. Please note that this clock and node hierarchy is mostly applicable to SONET/SDH environment. The first level of hierarchy is PRC that is defined by ITU-T G.811 and Telcordia

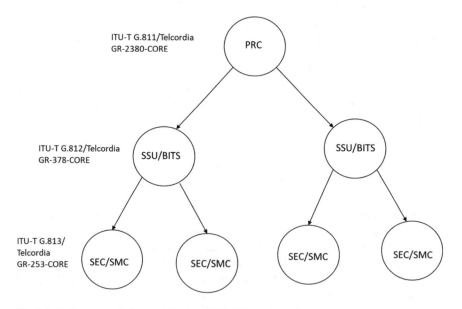

Fig. 3.2 A diagrammatical representation of PRC hierarchy

GR-2380-CORE. As discussed earlier a PRC can be either atomic frequency standard or UTC reference from GPS and high-quality OCXO.

The second level hierarchy is Synchronization Supply Unit (SSU) or Building Integrated Timing Supply (BITS) defined by ITU-T G.812 and Telcordia GR-378-CORE standards. The SSU or BITS includes holdover, a feature that allows it to generate a clock with higher accuracy than its intrinsic free running accuracy for a short period after it loses synchronization with PRC/PRS. SSU/BITS is usually implemented with a Digital PLL (DPLL) driven by a rubidium clock [3]. Third level in the hierarchy is SDH Equipment Clock (SEC) or SONET Minimum Clock (SMC) defined by ITU-T G.813 and Telcordia GR-253-CORE. These clocks also feature holdover, but its holdover and free-run accuracy performance is lower than those required in SSU/BITS clocks. The SEC/SMC clocks are usually implemented with a DPLL driven by OCXO or TCXO. It should be noted that PRC and SSU/BITS can be implemented as standalone but SEC/SMC are often implemented in a SONET/SDH add/drop multiplexer. Some of the telecommunications networks still use SONET/SDH infrastructure and thus SSU/BITS are considered existing install base for which newer deployments must provide backward compatibility or where applicable newer SSU/BITS equipment provides NTP, PTP synchronization options. Some of the T-GM and T-BC provide BITS interface for easier integration to BITS timing infrastructure. One of the key imperatives of PRC clock hierarchy deployment in SONET/SDH and TDM environment as defined in ITU-TG.811, G.812, and G.813 is node to node synchronization mechanism, core of which is frequency synchronization. Such node to node physical layer synchronization makes network sync more reliable, something that traditional telecom service providers prefer. One of the newest contenders in this technology is Synchronous Ethernet or SyncE. It uses frequency synchronization and needs node to node synchronization at PHY (Physical Layer) level. The PHY in a device implements physical layer functions and connects a link layer device to a physical medium, e.g., copper wire or fiber optics medium. We will explore further on SyncE in the proceeding chapter.

3.2.3   PRTC (Primary Reference Time Clock)

Defined by ITU-T G.8272/Y.1367, the primary Reference Time Clock (PRTC) is most common UTC reference clock deployed in modern networks. A PRTC provides time, phase, and frequency synchronization to a network or section of a network. A PRTC can be either implemented inside a T-GM or it can provide UTC input to T-GM as standalone device. However, in most cases T-GM integrates PRTC as integral part of the device. In some networks (e.g., smartgrid), a PRTC can be deployed as standalone mostly in the form of disciplined clock that provides up to 24 h holdover in case of a failure and implements high-quality OCXO as part of disciplined clock. WE will explore disciplined clock in proceeding chapter.

The PRTC provides reference time signal traceable to a recognized time standard (e.g., coordinated universal time (UTC)). In most common PRTC deployments, UTC is obtained from a global navigation satellite system (GNSS). However, UTC can also be obtained from a UTC time laboratory registered at Bureau International des Poids et Mesures (BIPM) (e.g., a national UTC time lab). An example of UTC time lab is NIST (National Institute of Standard and Technology) in the USA. Depending upon performance requirement, a network can deploy one of the two kinds of PRTC: PRTC-A and PRTC-B. The time output of PRTC-B is more accurate than that of PRTC-A. PRTC-B is intended for locations where it is possible to guarantee optimized environmental conditions (e.g., controlled temperature variation in indoor deployments). Examples are central location and large aggregation sites. PRTC-B is related to applications where more accurate absolute time is required. Achieving accurate relative time between timing chains may benefit from deploying PRTC-B clocks in all timing chains [4].

3.2.3.1   Time Error in Locked Mode

Under normal operating condition (device is not booting up or is in warmup condition), considered as locked mode, time output of PRTC-A or combined PRTC-A and T-GM functions must be within accuracy level of 100 ns or better when verified against primary time standard "UTC." This value of time error estimate includes all the noise components, i.e., the constant time error (time offset) and the phase error (wander and jitter) of the PRTC-A. For the combined PRTC-A and T-GM function, time error samples are measured through a moving-average low-pass filter of at least 100 consecutive time error samples. This filter is applied by the test equipment to remove errors caused by timestamp quantization, or any quantization of packet position in the test equipment, before calculating the maximum time error.

Similarly, under normal operating condition considered as locked mode as described above, the time output of PRTC-B or the combined PRTC-B and T-GM function must be accurate to within 40 ns or better when verified against the applicable primary time standard "UTC." This time error value includes all the noise components, i.e., the constant time error (time offset) and the phase error (wander and jitter) of the PRTC-B. In case of the combined PRTC-B and T-GM function, the time error samples are measured through a moving-average low-pass filter of at least 100 consecutive time error samples. As discussed above, this filter is applied by the test equipment to remove errors caused by timestamp quantization before calculating the maximum time error.

The ITUT-T G.8272 defines further timing parameters in terms of wander in locked mode expressed in MTIE. The following table depicts the wander generation measured in MTIE for PRTC-A and PRTC-B. This parameter is extremely useful for T-GM and PRTC designer and may be handy while calculating total phase noise for the network (Tables 3.5 and 3.6).

Table 3.5 Wander generation (MTIE) for PRTC-A [4]

MTIE limit [µs]	Observation interval τ [s]
$0.275 \times 10^{-3} \, \tau + 0.025$	$0.1 < \tau \leq 273$
0.10	$\tau > 273$

Table 3.6 Wander generation for PRTC-B

MTIE limit [µs]	Observation interval τ [s]
$0.275 \times 10^{-3} \, \tau + 0.025$	$0.1 < \tau \leq 54.5$
0.04	$\tau > 54.5$

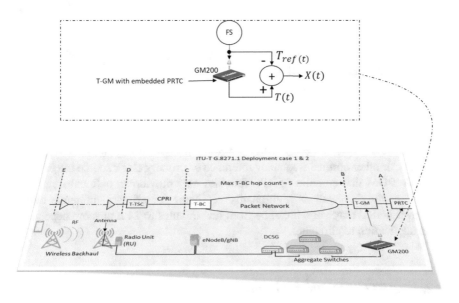

Fig. 3.3 Setup for PRTC-A and PRTC-B to measure MTIE and TDEV [5]

The setup for measuring MTIE and TDEV should be similar to those described in Chap. 2, Fig. 2.8. However, Fig. 2.8 of Chap. 2 is duplicated herein below for easier understanding of MTIE and TDEV measurement setup (Fig. 3.3).

In this diagram, Trimble's edge grandmaster (T-GM) GM200 is considered to depict the test setup. GM200 shown in top part of the diagram for test setup has built-in PRTC-A or better capabilities. Its accuracy level is around 50 ns little shy of the PRTC-B accuracy of 40 ns. ITU-T G.8272 defines the limit for wander expressed in TDEV for PRTC-A and PRTC-B as depicted in Tables 3.7 and 3.8.

Please note, MTIE and TDEV requirement for 1PPS interface are based on the time interval error of the 1PPS signal taken at one sample per second and without any low-pass filtering. The applicable requirements of MTIE and TDEV for an Ethernet interface carrying PTP messages are measured through a moving-average low-pass filter of at least 100 consecutive time error samples. This filter is applied

Table 3.7 Wander generation measured in TDEV for PRTC-A

TDEV limit [ns]	Observation interval τ [s]
3	$0.1 < \tau \leq 100$
$0.03\,\tau$	$100 < \tau \leq 1000$
30	$1000 < \tau \leq 10{,}000$

Table 3.8 Wander generation measured in TDEV for PRTC-B

TDEV limit [ns]	Observation interval τ [s]
1	$0.1 < \tau \leq 100$
$0.01\,\tau$	$100 < \tau \leq 500$
5	$500 < \tau \leq 100{,}000$

by the test equipment to remove errors caused by timestamp quantization before calculating the MTIE and TDEV.

3.2.3.2 Holdover

In case a PRTC loses its phase and time references, the PRTC must use its local oscillator or other optimal frequency reference or primary clock source traceable to UTC. A T-GM that combines PRTC function may provide an optional frequency input capability as well as high-quality oscillator for primary reference clock. Holdover period may vary, it can be from few minutes to 24 h or more depending upon application requirement.

3.2.4 PRTC Functional Model

A functional model of PRTC presented in ITU-T G.8272 defines interfaces and internal blocks of PRTC apparatus. However, the design and implementation of PRTC may vary vendor to vendor. A vendor may choose to follow its own design guideline and subsystem configuration for a PRTC apparatus and G.8272 PRTC functional model only serves as guideline, no bindings applied (Fig. 3.4).

The main purpose of a PRTC is to deliver the primary time reference to be used in time and/or phase synchronization of other clocks in the network. It receives a UTC time reference from a system having access to a recognized primary time standard (e.g., from a global navigation satellite system or from a national laboratory participating in time standards generation) and delivers this reference signal to all other clocks in a network. Additionally, a PRTC may include input and output frequency interfaces, but it must implement at least one output frequency interface. When connected to a frequency reference traceable to a PRC, the optional input frequency interface may be used to maintain the local representation of the timescale during outages of the input time reference (i.e., extend the phase/time holdover

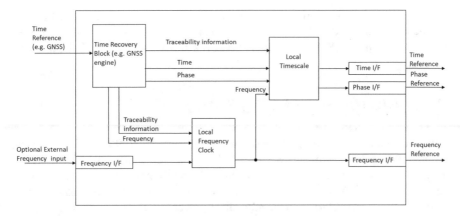

Fig. 3.4 The functional model of PRTC as defined in ITU-T G.8272 [4]

period of the clock). A possible use of the optional output frequency interface may be to measure the phase noise of the PRTC with traditional telecom signals [4].

The functionality of a PRTC defined by individual blocks is depicted in the diagram and can described as follows:

- *Time Recovery Block*: This block includes GNSS engine that receives and processes the external time interface such as from GNSS antenna and provides output signals to generate frequency, phase, and time. It also provides traceability information in process.
- *Local Time Scale*: The local time scale block maintains the local representation of the primary timescale, based on the frequency generated by the local frequency clock. It also generates the time and phase reference output signals.
- *Local Frequency Clock*: The local frequency clock generates the internally used frequency timing signals. In the case of GNSS signal loss by the time recovery engine, this clock may either go into holdover, or switch to the optional incoming frequency reference if such interface is present.
- *I/F*: Interface function necessary to generate a physical signal. An example of interface could be 1PPS or 10 MHz.

3.3 GNSS System Overview

We have discussed about importance of GNSS constellations to obtain primary reference clock traceable to UTC. In this section, we will learn about GNSS and other constellation as it applies to PNT (Positioning, Navigation, and Timing). As of writing this book, there are 6 GNSS Constellations proving PNT services across the globe. In addition, there are other possible satellite constellations that may be used for civilian use of PNT in future.

The 6 GNSS constellations providing PNT include GPS, GLONASS, Beidou, Galileo, IRNSS, and QZSS. Other constellations that provide similar services include Iridium, SBAS, and potentially SpaceX's newly launched Starlink satellites.

3.3.1   GPS (Global Positioning System)

First launched in the late 1970s, and operational since the 1990s, GPS is primary satellite constellations in the USA for PNT services and controller by US government. It remains the most popular constellation and offers full global coverage with approximately 32 medium Earth orbit satellites in six different orbital planes of the earth. Military and Civilian application share five bands, the most commonly used bands are 1.57542 GHz (L1 signal) and 1.2276 GHz (L2 signal). The modulation technique is CDMA. Recently as of 2009, the US Air Force successfully broadcast an experimental L5 signal on the GPS IIR-20(M) satellite. The first GPS IIF satellite with a full L5 transmitter launched in May 2010. In April 2014, the Air Force began broadcasting civil navigation (CNAV) messages on the L5 signal [6]. As of today, L5 is the third civilian GPS signal for PNT, designed to meet demanding requirements for safety-of-life transportation and other high-performance applications. It uses 1176.45 MHz for PNT applications. The GPS satellites fly in medium Earth orbit (MEO) at an altitude of approximately 20,200 km (12,550 miles) and each satellite circles the Earth twice a day. These satellites are arranged into six equally spaced orbital planes surrounding the Earth. Each plane contains four "slots" occupied by baseline satellites. This 24-slot arrangement ensures users can view at least four satellites from virtually any point on the planet. This arrangement is important consideration for PRTC since a GPS antenna connected to PRTC can get multiple satellites in view to triangulate and obtain traceable UTC for time distribution. As of writing this book, SpaceX in partnership with US Space Force's Space and Missile Center (SMC) has launched fourth GPS III satellite to Earth orbit on November 5th, 2020. Dubbed as "SV04," this new satellite is part of next generation GPS III program by SMC. This satellite is expected to be ready for operational use. It will join the constellation of 31 GPS satellites already on orbit in a few months. The GPS III satellites will provide three times greater accuracy and up to eight times improved anti-jamming capabilities over their predecessors. Once operational, this will be the 23rd M-Code-capable space vehicle in the GPS constellation, providing a more robust anti-jamming and anti-spoofing GPS signal to military users. The highly encrypted M-Code to protect GPS signals from jamming and spoofing currently is enabled on 22 GPS satellites of various generations; SV04 makes 23; 24 are needed to bring the M-Code to be fully operational [7].

The GPS III satellites features include:

- L1C signal on the 1575.42 MHz L1 frequency
- L2C signal on the 1227.6 MHz L2 frequency
- L5 "Safety of Life" signal on the 1176.45 MHz L5 frequency
- Military M-code

The GPS III constellation provides a new L1C civilian signal, which is interoperable with other GNSS system. The L1C signal shares the same center frequency as Europe's Galileo network, Japan's QZSS, and China's BeiDou.

3.3.2   GLONASS (Globalnaya Navigazionnaya Sputnikovaya Sistema, or Global Navigation Satellite System)

This satellites constellation is the second one launched in the late 1970s and operational since the 1990s. It is controlled by the Russian government. GLONASS constellation is considered second most popular GNSS for PNT and offers full coverage of the planet thanks to 24 medium Earth orbit satellites in three different orbital planes (better covering the poles than GPS) [8]. This constellation provides five shared bands for Military and civilian applications. The most commonly used bands are 1602 MHz (L1 signal), 1246 MHz (L2 signal), and 1202 MHz (L3). It uses FM modulation technique, but CDMA is being studied as well. Most GLONASS receivers also support GPS + GLONASS dual mode for enhanced operation. On October 25th, 2020 Russia launched its next generation Uragan-K No.15L satellite to join its GLONASS constellation. Each satellite in the GLONASS constellation is equipped with cesium atomic clocks to provide accurate times for its navigation signals.

3.3.3   BeiDou Navigation System (BDS)

It is the third group of satellites of GNSS constellation launched in 2000 and controller by China government. The BDS is operational since 2010 and second most popular in Asia. The Beidou system consists of three satellite constellations: Beidou-1, Beidou-2, and Beidou-3. The Beidou-1 consisted of three satellites launched in 2000 and offered limited coverage and PNT services, mainly for users in China and neighboring regions. It was decommissioned at the end of 2012. The second generation of BDS is Beidou-2 which was launched in 2009 and operational since December 2011. It had a constellation of 10 satellites. In 2015, China launched its third-generation satellite Beidou-3 on Earth's orbit and became operational on December 27th, 2018. The Beidou-3 or BDS-3 constellation has 35 satellites as of now, the last and final satellite was launched on June 23rd, 2020. This complete BDS-3 constellation provides global coverage for PNT functions.

3.3.4   Galileo

This constellation of satellites is Europe's own GNSS providing a highly accurate, guaranteed global positioning service under civilian control. The constellation will have 30 satellites to be fully commissioned but as of 2020 it has 26 live satellites on

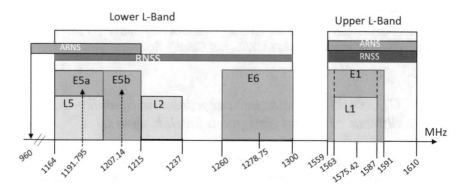

Fig. 3.5 Galileo E1, E5a, E5b, and E6 bands and corresponding GPS L1, L2, and L5 bands (courtesy: esa) [9]

orbit and 22 are operational. The program is expected to be complete and fully operational by 2021. First Galileo satellite "GIOVE-A" was launched on 28 December 2005. Once fully operational, the constellation will offer global coverage using up to 30 medium Earth orbit satellites in three different orbital planes. Publicly regulated and civilian applications share four bands, the most commonly used being 1575.420 MHz (E1) and 1176.45 MHz (E5a). The modulation technique is CDMA. Galileo receivers will also support GPS + Galileo dual constellation mode for enhanced operation.

Figure 3.5 shows Galileo and GPS corresponding bands. Galileo offer E1, E5a, E5b and E6 bands. The Galileo frequency bands have been selected in the allocated spectrum for Radio Navigation Satellite Services (RNSS) and in addition to that, E5a, E5b, and E1 bands are included in the allocated spectrum for Aeronautical Radio Navigation Services (ARNS), employed by civil aviation users and allowing dedicated safety-critical applications [9].

The Galileo constellation has passive hydrogen maser clock as the master clock on board the spacecrafts providing atomic frequency standard with accuracy of 0.45 ns over 12 h. A rubidium clock will be used should the maser clock fail. It is accurate to within 1.8 ns over 12 h.

3.3.5 Indian Regional Navigation Satellite System (IRNSS)

The IRNSS with an operational name of *NavIC* (acronym for Navigation with Indian Constellation) is an independent regional navigation satellite system being developed by India. The first satellite of NavIC constellation was launched on July 1st, 2013 and operational since 2018. NavIC constellation has eight satellites and provides accurate position information service to users in Indian region extending up to 1500 km from India's boundary, which is its primary service area. An extended service area lies between primary service area and area enclosed by the rectangle from Latitude 30° South to 50° North, Longitude 30° East to 130° East [10]. It offers 2 shared bands for military and civilian applications: 1176 MHz (L5 signal) and 2492 MHz (S signal). The modulation technique is CDMA.

3.3.6 Quasi-Zenith Satellite System (QZSS)

The QZSS is a Japanese satellite positioning system composed mainly of satellites in quasi-zenith orbits (QZO). However, the term "Quasi-Zenith Satellite (QZS)" can refer to both satellites in QZO and geostationary orbits (GEO). The constellation objective is to broadcast GPS-interoperable and augmentation signals as well as original Japanese (QZSS) signals from a three-spacecraft constellation in inclined, elliptical geosynchronous orbits. It was first launched in 2010 and operational since 2018. The constellation currently has four satellites and plan to have seven in total. It offers local coverage with one geostationary satellite and three satellites in quasi-zenith orbit above Japan, Southeast Asia, and Australia. Both military and civilian applications share the same bands and modulation technique as GPS. To reduce errors in satellite positioning, QZSS provides Sub-meter Level Augmentation Service (SLAS) in the L1S signal and Centimeter Level Augmentation Service (CLAS) in the L6 signal. In the future, QZSS will offer authentication messages that will be provided with navigational messages as measures against spoofing attacks.

3.3.7 Iridium Constellation

Iridium is not part of GNSS constellation, but this constellation of satellite also provides PNT services for military and civilian use. This constellation has 66 satellites that are connected in a cross-linked mesh architecture providing global coverage. In 2007, Iridium Satellite LLC announced its plans to develop its Iridium NEXT constellation and start deployment in the timeframe 2015–2017. The first generation of Iridium satellites are developed by Iridium communications Inc. and financed by Motorola. These satellites were deployed in 1997 to 2002. The first test telephone call was made over the network in 1998, and full global coverage was completed by 2002. Later, Iridium communications began to deploy its second-generation Iridium NEXT into the existing constellation in January 2017. The Iridium NEXT satellite constellation has the capability in its L-band service links to provide mobile satellite service in 1616.0–1626.5 GHz using Time Division Duplex (TDD) between the uplink and the downlink signals [11].

3.3.8   SBAS (Satellite-Based Augmentation System)

The SBAS satellites augment primary GNSS constellation(s) by providing GEO ranging, integrity, and correction information. While the main goal of SBAS is to provide integrity assurance, it also increases the accuracy with position errors below 1 meter (1 sigma). Its GEO ranging feature provides transmission of GPS-like L1 signals from GEO satellites to augment the number of navigation satellites available to the users. The signals are bi-phase shift key (BPSK) modulated by a bit train comprising the PRN code and the SBAS data (modulo-2 sum). The following is a list of SBAS services:

- *Wide Area Augmentation System (WAAS)*: Developed by FAA (Federal Aviation Administration), WAAS provides GPS corrections and a certified level of integrity to the aviation industry, to enable aircraft to conduct precision approaches to airports. The corrections are also available free of charge to civilian users in North America.
- *European Geostationary Navigation Overlay Service (EGNOS)*: The EGNOS is developed by ESA (European Space Agency) in cooperation with the European Commission (EC) and EUROCONTROL (European Organization for the Safety of Air Navigation). It improves the accuracy of positions derived from GPS signals and alerts users about the reliability of the GPS signals. There are three EGNOS satellites that covers European Union member nations and several other countries in Europe. The constellation transmits differential correction data for public use and has been certified for safety-of-life applications.
- *MTSAT Satellite Based Augmentation Navigation System (MSAS)*: This SBAS constellation provides augmentation services to Japan. It uses two multifunctional transport satellites (MTSAT) and a network of ground stations to augment GPS signals in Japan.
- *GPS-Aided GEO Augmented Navigation System (GAGAN)*: The GAGAN supports flight navigation over Indian airspace. It is based on three geostationary satellites, 15 reference stations installed throughout India, three uplink stations, and two control centers. GAGAN constellation is compatible with other SBAS systems, such as WAAS, EGNOS, and MSAS.
- *System for Differential Corrections and Monitoring (SDCM)*: The SDCM is developed by Russia to provide Russia with accuracy improvements and integrity monitoring for both the GLONASS and GPS navigation systems. Space Segment includes three operating geostationary satellites of multifunctional Space System Launch, broadcasting SDCM data to users by means of SBAS radio signals.
- *Other SBAS Systems:* There are few other countries also taken initiative to launch their own SBAS constellation, e.g., China and Korea.

References

1. Raltron. (n.d.) *Stratum levels defined*. Product and Technology Application Notes. Raltron Electronics Corporation.
2. Dryburgh, L., & Hewett, J. (2005). *Signaling system no. 7 (SS7/C7): protocol, architecture, and services*. Cisco Press.
3. Milijevic, S. (2009). *An introduction to synchronous ethernet*. Embedded by Aspen Core. Available online at https://www.embedded.com/an-introduction-to-synchronized-ethernet/
4. G.8272. (2018). *Timing characteristics of primary reference time clocks*. ITU-T G.8272/Y.1367 (11/2018). International Telecommunication Union.
5. G.8271.1. (2020). *Network limits for time synchronization in packet networks with full timing support from the network*. ITU-T G.8271.1/Y.1366.1. International Telecommunication Union.
6. GPS.gov. (2020). *New civil signals*. GPS.GOV. Available online at In April 2014, the Air Force began broadcasting civil navigation (CNAV) messages on the L5 signal.
7. Dhande, M. (2020). *What is GPS III & how capable it is?* Geospatial Media and Communications. Available online at https://www.geospatialworld.net/blogs/what-is-gps-iii-and-what-are-us-gps-iii-capabilities/
8. GSTR-GNSS. (2020). *Consideration on the use of GNSS as a primary time reference clock in telecommunications*. ITU-T GSTR-GNSS. International Telecommunication Union.
9. esa. (2020). *Galileo signal plan*. European Space Agency. Available online at https://gssc.esa.int/navipedia/index.php/Galileo_Signal_Plan
10. ISRO. (2020). *Indian regional navigation satellite system (IRNSS): NavIC*. Department of Space, Indian Space Research Organisation. Available online at https://www.isro.gov.in/irnss-programme
11. Iridium. (n.d.) *Iridium NEXT engineering statement*. Iridium Constellation LLC.

Chapter 4
GNSS Time

4.1 Introduction

GNSS-based PRTC is a common source of PRS for network wide time distribution. This solution is very cost-effective to be integrated in routers and switches, T-GM, T-BC, and T-TSC devices make it easier to create effective time distribution across the network. Although it may not be as accurate and stable as an atomic clock, given its nanosecond-level accuracy and cost GNSS-based PRTC is highly effective in critical infrastructure environment. Moreover, the stability of clock can be further ascertained through built-in oscillator in a PRTC device. However, there are multiple factors that may render time errors in GNSS receivers.

For this purpose, it is important to learn about how GNSS system operates, keeps its time synchronized and offers PNT services. More importantly, how PRTC derived UTC traceable time from GNSS System. Although GNSS system operates with precise timekeeping mechanisms, multiple sources of errors could degrade the signal before it is received and processed by the PRTC for time synchronization. Knowing about these errors and how to mitigate them either at PRTC level or at network level will be handy in designing distributed precision time synchronization across a critical network infrastructure.

In this chapter, we will explore GNSS system including its internal subsystems and their operation, GNSS Time, errors that impact GNSS time accuracy and GNSS signals and how to mitigate them. Readers will find learning in this chapter is very useful while selecting a PRTC product to design their network infrastructure.

© The Author(s), under exclusive license to Springer Nature
Switzerland AG 2021
D. D. Chowdhury, *NextGen Network Synchronization*,
https://doi.org/10.1007/978-3-030-71179-5_4

4.2 GNSS System

In Chap. 3, we explored various GNSS constellations that offer PNT services across the globe and some uses onboard atomic frequency standard for time distribution. In this section, we will further explore how GNSS time is obtained by a PRTC device. All GNSS satellite systems can be divided in three segments from operational perspectives:

Control Segment This segment consists of a coordinated, hierarchical group of stations on various locations on the ground. For example, GPS has 29 ground stations across the globe. Figure 4.1 illustrates global GPS ground stations for control segment operations.

The GPS operational control segment (OCS) includes 1 master control station, an alternate master control station, 11 command and control antennas, and 16 monitoring sites. These ground stations include ground atomic clocks acting as master clock to the clocks within satellites, as well as controlling or monitoring links to all satellites. The QZSS block diagram in Fig. 4.2 depicts how atomic clock reference is distributed as reference clock for satellite's internal system and carrier wave.

The onboard atomic clock also known as "space atomic clock" must meet stringent requirements from launching to unattended operation for many years. These clocks must assure satisfactory and reliable performances over overall mission life, meet constraints on mass, volume, and power consumption, survive launch environment (shock, acceleration, and vibration) and survive operational environment (vacuum, thermal cycling, EMI/EMC, radiation, magnetic field, and other space hazards) [3]. Table 4.1 shows different space atomic clocks for GNSS constellations.

Fig. 4.1 Network of GPS global ground stations for control segment operations (Courtesy: GPS. gov) [1]

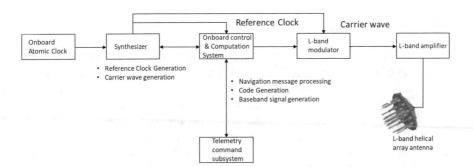

Fig. 4.2 The diagrammatical representation of GNSS satellite with onboard atomic clock. This block diagram is specifically related to QZSS [2]

Table 4.1 GNSS constellations and their respective onboard space atomic clock [3]

GPS	GLONASS	GALILEO	BEIDOU	IRNSS	QZSS
Rubidium	Cesium	Hydrogen maser	Rubidium	Rubidium	Rubidium
Cesium		Rubidium			

Apart from clock reference standards, the control segment is responsible for many other operations such as satellite enabling, orbit or time corrections, or uploading useful data (such as ephemeris and ionospheric model).

Space Segment This segment consists of the satellites, under the management of the control segment, and relaying information broadly (time, ephemeris, status) to all receivers located on the surface of the Earth, in other words space segment is responsible for transmitting radio signals to users. The satellites include an energy source to allow multi-channel radio transmissions to and from the control and user segments, as well as orbit adjustments. They also include high stability oscillators (atomic clocks on most constellations) that can be tuned by the control segment to keep optimal time accuracy [4].

User Segment The user segment is made of millions of GNSS receivers, some of them mobile (smartphones or vehicles, for example), some of them static (differential GPS references or telecom clocks, for example). Each one is designed for a low-cost recovery of its location and time offset, while receiving the signals from multiple satellites. Some receivers allow dual band reception for one constellation, or even for multiple constellations [4].

4.2.1   GNSS Time

GNSS system uses various timescale and parameters for processing time. The following diagram shows timescale and parameters in GPS system and how the time
information is processed in the GPS signal. This is applicable to GNSS system as
most GNSS system has similar mechanism for time processing. With the onboard
atomic clock, a GNSS system uses its own time, for GPS system it is known as
"GPS Time" or "GPS System Time Scale." This GPS time is established by the
control segment and is used as the primary time reference for all GPS operations.
The GPS time is referenced to a UTC zero time-point maintained by the U.S. Naval
Observatory (USNO). The zero time-point defined as midnight on the night of
January 5, 1980 (morning of January 6, 1980). At that epoch, the difference between
TAI and UTC was 19 s, hence GPS − UTC = n − 19 s. The GPS time is synchro-
nized with the UTC (USNO) at 1 microsecond level (modulo 1 s), but actually is
kept within 25 ns [5].

   As illustrated in Fig. 4.3, each satellite in GPS constellation maintains its internal time scale, named SV (Space Vehicle) time which receives the updated models
of the differences SV time versus GPS time, and GPS time versus UTC (USNO)
from the ground monitoring station as ground segment by uplink channel. This

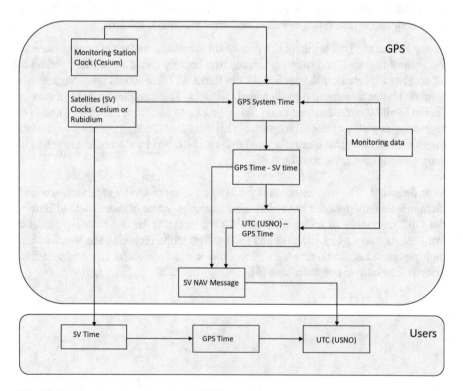

Fig. 4.3 Time scales and parameters in GPS System [4]

modelling information is present as various parameters in GPS navigation messages embedded in each emitted GPS electromagnetic signal towards the Earth. A GPS receiver must first track a GPS satellite or space vehicle (SV) on one or both of the L-band carriers to set its time to a GPS time. After locking on to the spread spectrum carrier by applying the local pseudorandom noise (PRN) code at the correct time, the 50-hertz data stream containing the navigation message can be decoded. The time as given by the individual satellite's clock (SV time) can be recovered by noting the star time of the PRN code necessary to achieve lock. GPS receiver then processes all these parameters from all received satellite signals to rebuild a local realization of GPS or UTC (USNO) time scale and produce a physical timing signal on a PPS output port.

Other timescale such as GLONASS (GLONASST) is generated by the GLONASS central synchronizer (CS) and the difference between the UTC (SU) and GLONASST should not exceed 1 ms plus 3 h. The CS consists of an ensemble of hydrogen clocks (HC) having an instability of $(1...3) \times 10 - 14$ (averaging interval 24 h). Similarly, a continuous time scale maintained by the Galileo Central Segment and synchronized with Galileo System Time (GST) TAI with a nominal offset below 50 ns. The GST that runs on Galileo constellation is under responsibility of the Galileo Mission Segment (GMS) and fixed on the ground at the Galileo Control Centre in Fucino (Italy) by the Precise Timing Facility (PTF), based on the average of different atomic clocks. The GST start epoch is 00:00 UT on Sunday 22nd August 1999 (midnight between 21st and 22nd August). At the start epoch, GST was ahead of UTC by 13 leap seconds. Since then 3 additional leap seconds have been introduced (31 December 2005 and 2008, and 30 June 2012). Therefore, currently GST is ahead of UTC by 16 s [6].

The BDT (Beidou System Time) is generated by ensembles of master control station (MCS) and monitor station (MS). Beidou's control and ground segment consists of a MCS, two upload station, and a network of 30 widely distributed MS. The MCS is responsible for the operational control of the system, including orbit determination, navigation messages, and ephemerides which are based on the China Geodetic Coordinate System 2000 (CGCS2000). It also coordinates mission planning, scheduling, and time synchronization with BeiDou Time (BDT). BDT is synchronized within 100 ns of UTC (NTSC) as maintained by National Time Service Center, China Academy of Science. BDT does not incorporate leap seconds. The leap second offset is broadcasted by the BeiDou satellites in the navigation message. The initial epoch of BDT is 00:00:00 UTC on January 1, 2006. The offset between BDT and GPST/GST is also to be measured and broadcasted in the navigation message [7].

Table 4.2 depicts GNSS time references for GPS, GLONASS, Galileo, and Beidou.

Table 4.2 GNSS Time References

Constellation	GPS	GLONASS	Galileo	Beidou
Launch	First launch 1978 Fully operational 1995	First launch 1982 Fully operational 2011	First launch 2011 24 operational satellite as of 2020	First launch 2000 Fully operational 2020
Constellation time format	*GPS time* Continuous timescale, starting on January 6, 1980	*GLONASS time* UTC (SU), generated by state time and Frequency reference (STFR).	*Galileo time* Continuous timescale, starting on august 22, 1999	*BeiDou time* Continuous timescale, starting on January 1, 2006
Standard time format	UTC (USNO)	UTC (SU)	TAI	UTC (NTSC)
Relation with UTC	UTC = GPS ± leapsecond	GLONASS time = UTC (SU) + 3 h	UTC = GST + Δ seconds	BDT = UTC (NTSC)

4.2.2   Sources of Time Error in GNSS Time Distribution

Although how GNSS constellations perform timekeeping and synchronized their respective time may vary slightly, obtaining timescale by GNSS receiver is similar to how GPS time is processed by a GPS receiver as shown in Fig. 4.3. All GNSS receivers obtain GNSS time or UTC timescale by receiving space signals into a GNSS receiver. However, GNSS signals may encounter some errors before it is received and processed by the receiver. These errors on GPS/GNSS signal transmission must be corrected in order to obtain high-precision satellite clock time. As shown in Fig. 4.4, GNSS signals are encountering different errors as it enters Earth's atmosphere and processed by the receiver. For the simplicity of understanding, we classify these errors into three categories: satellite-related errors, signal propagation-related errors, and receiver-related errors. According to statistical calculations, the accuracy of one-way timing can be as much as 20 ns after correcting for such errors.

4.2.2.1   Satellite-Related Error

These types of errors are often known as "satellite ephemeris errors" which are related to deviation between the calculated satellite position and the actual position of the satellite. Since GNSS receiver uses the satellite's location in position calculations, an ephemeris error reduces user accuracy. Another satellite-related error is satellite clock error that occurs due to frequency offset, frequency drift, and other effects constituting variation between the onboard atomic clocks and GNSS standard time. Although onboard atomic clocks are highly stable, they lack perfect synchronization between the timing of the satellite broadcast signals and GPS/GNSS system time. The satellite clock error is caused by oscillator not being synchronized

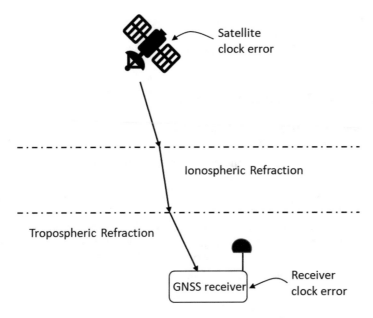

Fig. 4.4 Sources of time error in GNSS time distribution

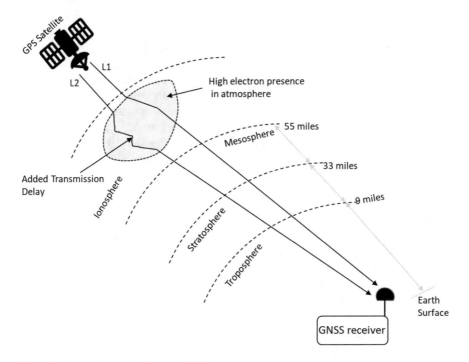

Fig. 4.5 Propagation-related error of GPS signals

to the GPS/GNSS time and it is one of the main errors which affect the positioning accuracy. The satellite clock error will cause an error in the pseudorange, resulting in a time error for a GNSS receiver's local time.

4.2.2.2 Signal Propagation-Related Error

The GPS/GNSS signal passes through near-vacuum of space and then through atmosphere of Earth. The following diagram shows that two GPS satellite signals L1 and L2 entering Earth's atmosphere are acquiring transmission delay due to radio signals getting bent as it passes through atmosphere.

To obtain accurate position and time, GPS/GNSS receiver needs to know the length of the direct path from the satellite to the GNSS receiver. But, as shown in Fig. 4.5 radio signals are bending as they enter the Earth's atmosphere. This bending increases the amount of time the signal takes to travel from the satellite to the receiver. The first delay it encounters is the ionospheric delay. For the discussion here, we will explore ionospheric and tropospheric delay.

4.2.2.3 Ionospheric Delay

The layer of the atmosphere that most influences the transmission of GPS/GNSS signals is the ionosphere, the layer is approximately 55 mile above the Earth's surface. The ionosphere is ionized plasma comprised of negatively charged electrons which remain free for long periods before being captured by positive ions. It is the first part of the atmosphere that the signal encounters as it leaves the satellite. The magnitude of the delay induced by ionosphere depends on the state of the ionosphere during the moment the signal passes through. Its density and stratification vary and are influenced by the Sun. When gas molecules at ionosphere are ionized by the Sun's ultraviolet radiation, free electrons are released. As their number and dispersion varies, so does the electron density. The higher the electron density, the larger the delay of the signal, but the delay is by no means constant. This delay could be range from few nanoseconds to 25 ns for GNSS signal [4].

ITU-T technical report "GSTR-GNSS" suggested few alternatives to address the uncertainty of ionospheric delay. First, correct ionospheric delay using a model. Single frequency GNSS receivers can use a model to provide an approximate correction for ionospheric delay. For example, most single band GPS receivers apply a correction using the Klobuchar model [8]. All GPS satellites broadcast parameters of this model in the GPS navigation message. This model typically reduces the time error due to ionospheric delay by 50%.

Secondly, use regional correction services like SBAS that provide improved accuracy of the ionosphere delay model. Third, mitigate with primary atomic clocks or disciplined clock (we will introduce disciplined clock in the proceeding chapters). This is an effective strategy that utilizes diurnal filtering and space weather detection to prevent adjusting the local atomic timescale to incorrect time. This strategy can be used effectively in ePRTC defined by ITU-T G.8272.1 and can be used either in conjunction with multiband receivers or with single frequency receivers [4]. Fourth, user may utilize multiband GNSS receivers to actively compensate for atmospheric time delay.

4.2.2.4   Tropospheric Delay

This type of delay added by lower atmospheric condition. It is normally calculated as a function of local temperature, pressure, relative humidity, and the elevation angle of the satellite. For GPS Signals, both L1 and L2 are equally delayed, so the effect of tropospheric delay cannot be eliminated the way ionospheric delay can be. It is possible, however, to model the troposphere, then predict and compensate for much of the delay. For example, the effect on the timing of tropospheric delay of zenithal propagation is about 7.59 ns and the effect of tropospheric delay at an elevation angle of 10° is about 66 ns for GPS signal. If the meteorological parameters such as temperature, humidity, barometric pressure, and vapor pressure are used to correct the tropospheric delay in the observation station, the impact on timing can be reduced to sub-nanosecond [4].

4.2.2.5   Multipath Effects

GNSS signal may reflect by obstacles near the GNSS receiver's antenna before received by the receiver. These obstacles could be a building or the reflective materials at any structure behaving as mirrors causing multiple delayed reflected signals interfering with the direct signal from the satellite. This multiple reflection or otherwise known as "multipath" causes deviation between the observed time value and the direct signal. Multipath is a major source of error and affects accuracy of GNSS time estimation. Multipath is difficult to correct by a model and it affects both LOS and NLOS (Non-Line-Of-Sight) satellite signals. Figure 4.6 illustrates LOS (Line of Sight) scenario in which multipath signals accompany direct signals. The peak timing position of the correlator output is almost the same as that for the direct signal, so time synchronization accuracy is barely affected.

In the LOS scenarios, there are signal processing techniques to isolate the direct signal from the reflected signal. Whereas NLOS case, multipath signals from non-visible (NLOS) satellites are without accompanying direct signal as shown in Fig. 4.7. The peak timing position of correlator output is different from that of the direct signal due to propagation delay, with a large effect on time synchronization accuracy. In case of LOS, direct signal is stronger and received first before weaker reflected signals received by the receiver. Hence, the effects can be cancelled effectively through the signal processing mechanisms of the correlator circuitry in the GNSS receiver. However, since there is no direct signal in the case of NLOS, filtering through correlator circuitry is not be effective. The best course of action should

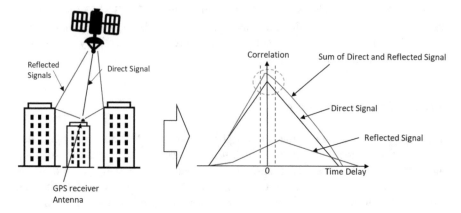

Fig. 4.6 The multipath issue in Line Of Sight (LOS) scenario

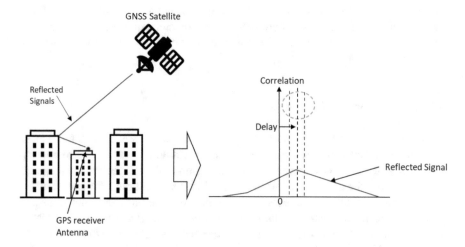

Fig. 4.7 The multipath issue in NLOS scenario

be to avoid NLOS scenarios through positioning and repositioning of GNSS receiver antenna until a direct path is achieved between the antenna and the GNSS satellite.

4.2.2.6 Multiple Antenna Cable Reflections

We discussed about multipath and its effect in earlier section. These reflected GNSS signals can enter at the receiver due to imperfect impedance matching at the receiver and antenna. Also, it may bounce from the receiver to the antenna and back to the receiver, transiting the antenna cable two times more than the direct signal as shown in Fig. 4.8.

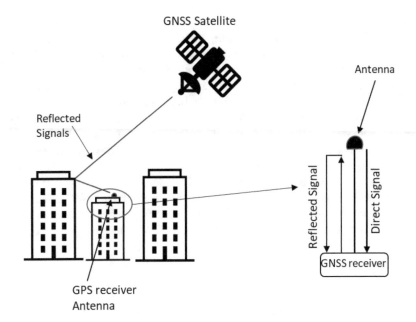

Fig. 4.8 Multiple antenna cable reflection

Henceforth, it is important that the impedance matching of the antenna cable with the antenna and receiver terminations be controlled to limit the power of a reflected signal.

4.2.3   Mitigation of Time Error in a PRTC

According to ITU-T technical report [4], the accuracy of one-way timing can be as much as 20 ns after correcting for these three categories of errors discussed above. A GNSS receiver can implement several mitigation methods to eliminate some of the errors and those are: SNR masking, Elevation Masking, P/TDOP, T-RAIM, and cable delay correction. These parameters can be used to configure receiver to ignore satellites that are expected to deteriorate the performance.

4.2.3.1   SNR (Signal to Noise Ratio) Masking

The SNR is a measure of the information content of a GNSS signal relative to the signal's noise, or the ratio of good information to degraded information. This quality information of GNSS signal is often reported by GNSS receiver as carrier-to-noise power ratio (C/N_0) values. A low C/N_0 values can result from low-elevation satellites, partially obscured signals (due to dense foliage for example), or reflected RF signals (multipath). A value above 20 is very good (20 parts good for every one part degraded). The quality of a position is degraded if the signal strength of any satellite in the constellation is below 6. Urban canyon is a good example of situation in which satellite signal can be degraded due to multipath. This happens due to signals getting reflected and contaminated by many tall buildings and the preponderance of reflective materials in urban environment. This combination of direct and reflected signals at the receiver tend to have low C/N_0 values, since the multipath reflected signals mask the true shape of the original signal and hence make it difficult for the receiver to detect it in the noise. Another example of low C/N_0 values can occur due to similar non-line-of-sight (NLOS) signals without direct signals. Apart from urban canyon, NLOS can also result from any other objects blocking direct view of the sky.

With the help of C/N_0 values, an engineer can determine if GNSS antenna positioning is correct and whether the antenna is getting correct view of the sky. Repositioning of the antenna may result in higher C/N_0 values in some cases.

4.2.3.2   Elevation Masking

High elevation satellites have better quality signals than lower elevation satellites. The elevation mask stops the receiver from using satellites that are on low elevation. Atmospheric errors (such as refraction from ionospheric and tropospheric layers) and signal multipath are largest for low-elevation satellites. Rather than attempting to use all satellites in view, a GNSS receiver should exclude satellites with very low elevation from GNSS fix and timing computations. This process can reduce the likelihood of the potential errors discussed above.

4.2.3.3   P/TDOP

The Position/Time Dilution of Precision (P/TDOP) is a measure of the error caused by the geometric relationship of the satellites used in the position solution. A low DOP (Dilution of Precision) value of 4 or less indicates good satellite geometry, whereas higher DOP values indicate that satellite geometry is weak. Table 4.3 depicts corresponding rating for DOP values.

4.2.3.4   T-RAIM Masking

The Time-Receiver Autonomous Integrity Monitoring (T-RAIM) is an algorithm of checking the timing solution integrity. It allows computing the distribution of time recovery from all visible individual satellites, and to reject the outliers. This is the timing application equivalent to the RAIM algorithms used for positioning applications. Once enabled, it will mask off signals from satellites that indicate any fault conditions, or if the particular GNSS signal is outside certain limits compared to three or more other satellite signals.

Table 4.3 DOP values and corresponding rating [9]

DOP value	Rating	Description
1	Ideal	Highest confidence level for applications that demand highest precision at all times
1–2	Excellent	Accurate enough confidence level to meet sensitive application needs
2–5	Good	Minimum appropriate for making accurate decision
5–10	Moderate	Can use for calculation, fix quality needs to be improved
10–20	Fair	Low confidence level
>20	Poor	Inaccurate measurement

4.2.3.5 Cable Delay Correction

Since position estimated by receiver is that of the antenna, cable length between the antenna and receiver and all electronics circuitry therein including that of the receiver contributes to adding a delay in the reception process. This total delay will bias the output 1PPS to "too late" to arrive at user application. These delays cannot be solved by the equation by the receiver since they are common for all satellites. If it is not manually configured, this error cannot be eliminated and will add the time error to user application. It is therefore recommended that the user configurable interface offers a compensation parameter for these various delays. It is possible to manage this compensation by adjusting time of deliverance of the 1PPS pulse, or by adjusting the data that says what time the pulse arrives.

4.2.3.6 Methods of Time Error Measurement in Multipath Environment

The multipath scenario we discussed as sources of errors is often found in urban canyon environment where multiple tall buildings may obscure direct path to GNSS signals. A 3D model may be created to map the urban environment and then a 3D ray-trace simulator used to calculate all the possible paths from a GNSS satellite to the reception point. Such ray-tracing model will help determine where the direct signal is obscured, and where it is reflected or diffracted by buildings or other objects.

Figure 4.9 shows GNSS satellite data is passed to 3D ray-trace simulator and ray-tracing model therein it is used as input to the GNSS simulator to enable it to generate both the direct and reflected or diffracted signals.

Later, the GNSS simulator will generate multiple signal replicas per satellite to reproduce the direct and multipath signals based on to calculated multipath profile. For testing PRTCs, the GNSS simulator needs to have a frequency reference of better stability than the PRTC itself. In this experiment, Trimble's GM200 is acting both as PRTC and T-GM.

The 1PPS output of the simulator is phase locked to the incoming frequency reference and aligned to the 1 s boundary of the RF signal before impairment by the multipath model. The manufacturer of the GNSS simulator should ascertain the accuracy of RF to 1PPS and it is recommended that the equipment is periodically

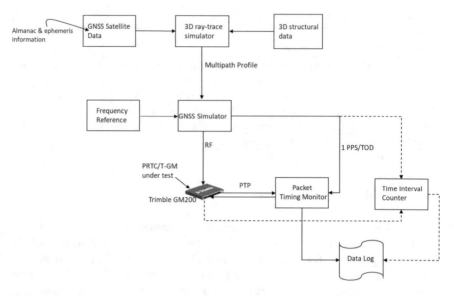

Fig. 4.9 Method of PRTC time error measurement in a multipath environment

calibrated. The ToD associated with the 1PPS from the GNSS simulator is not required to do the time error measurement. It is used to verify the PTP second [4].

References

1. gps.gov. (2018). *Control segment*. GPS.GOV. U.S. Air Force. Available online at https://www. gps.gov/systems/gps/control/
2. esa. (2020). *QZSS (Quasi Zenith Satellite System)*. European Space Agency (ESA). Available online at https://earth.esa.int/web/eoportal/satellite-missions/q/qzss
3. Rochat, P., Droz, F., Wang, Q., & Froidevaux, S. (2012). Atomic clocks and timing systems in global navigation satellite systems. In: *the European navigation conference*, 25–27 April, Gdansk (Poland). Available at ResearchGate.
4. GSTR-GNSS. (2020). *Considerations on the use of GNSS as a primary time reference in telecommunications* (ITU-T Technical Report). International Telecommunication Union.
5. IS-GPS-200. (2012). *Global positioning systems directorate systems engineering & integration interface specification (IS-GPS-200G)*. Navstar GPS Space Segment/Navigation User Interface. GPS Directorate, Revision G, 13 January 2013.
6. Galileo GNSS. (n.d.). *GST*. Galileo System Time. Galileo GNSS. Available online at https:// galileognss.eu/gst-galileo-system-time/
7. Sickle, V. J. (n.d.). *Chinese BeiDou*. Department of Geography, PennState College of Earth and Mineral Science. Available online at https://www.e-education.psu.edu/geog862/node/1879
8. Klobuchar, J. (1987). Ionospheric time-delay algorithms for single-frequency GPS users. *IEEE Transactions on Aerospace and Electronic Systems, 3*, 325–331.
9. Wikipedia. (2020). *Meaning of DOP values*. Wikipedia. Available online at https://en.wikipedia. org/wiki/Dilution_of_precision_(navigation)

Chapter 5
Timing Devices

5.1 Introduction

Delivering time synchronization throughout the network requires careful formulation to keep overall error budget small so that time and phase alignment is optimized through the network. The critical element of this preparation is the selection of timing devices that works in tandem to distribute time and phase alignment through the network. In a packet network, timing devices need to work with the required asymmetry compensation since packet network is subject to asymmetric condition such as delay and jitter, etc. For example, a boundary clock must accommodate asymmetry condition to provide master clock downstream. Thus, understanding the function of various timing devices, their performance expectations and error budget considerations are essential to design an optimized network for time-sensitive traffic such as 5G fronthaul. In this chapter, we introduce GRU (GNSS receiver unit) that provides GNSS timing reference to master clocks of the network. Understanding GRU functions is important as it is critical element of packet timing device such as grandmaster clock. A GRU can work in standalone mode providing PRTC to packet timing devices or can be integrated as part of it. A variation of GRU known as disciplined clock provides stable clock with accuracy compared to atomic clock and provides longer holdover time in case of GNSS signal failure. The disciplined clock is a cost-effective alternative to atomic clock reference. The concept of PRTC and GRU is further expended with introduction packet timing device depicting how GRU is playing critical rule for each. Later, various packet timing devices and their operational parameters are explored. The concept of GRU and packet timing devices presented in this chapter will help readership to select appropriate timing devices to create frequency, phase, and time alignment throughout the network.

D. D. Chowdhury, *NextGen Network Synchronization*, https://doi.org/10.1007/978-3-030-71179-5_5

5.2 GNSS Receiver Unit (GRU)

In discussing GNSS time at Chap. 4, we explored three key elements of GPS/GNSS time: control segment, space segment, and user segment. In this section, we will explore further on user segment that mainly comprises GPS/GNSS receiver that is designed to passively acquire and track GPS/GNSS signals. The following diagram depicts an external GNSS Receiver Unit (GRU). The GNSS receiver chip is an SOC (System on Chip) and readily available to the market from a number of vendors. The SOC can be GPS or GNSS solution that includes CMOS RF, baseband, and a micro-controller subsystem (Fig. 5.1).

The microcontroller subsystem generally includes a CPU, FPU, cache controller, memory management, and protection unit. The GNSS receiver SOC is presented in Fig. 5.2.

The 32 bits bus matrix connects all subsystems including microcontroller, GNSS Receiver RF and baseband unit, Crypto engine, DMA and SPM and I/O. Depending upon clock accuracy needed, an appropriate quartz oscillator such as TCXO or OCXO is connected to dynamic clock management. The antenna circuitry includes LNA (low noise amplifier), filters, and surge protection circuit. Because of the extremely low powered received signals on the order of −160 dBw, the GNSS receiver is particularly susceptible to interference and by design, have a high sensi-tivity and low SNR. It is often difficult to obtain proper reception in urban environ-ments that are subject to interference and obstacles. These anomalies cause multipath and fading effects as discussed in Chap. 4. The filtering is thus vital for rejecting out-of-band signals, reducing the noise, mitigating the impact of aliasing, and pro-viding burnout protection by removing high power RFI found within multi-located base station environments.

The function of GRU is to provide PRTC function with 1PPS, TOD, and 10 MHz inputs to systems connected to it. In standard telecom environment, a GRU is con-nected as an external device to cell site router providing clock input traceable to UTC. It synchronizes its clock with GNSS time and provides input to traceable UTC clock input to devices connected to it. In traditional telecom network design, a cell site router implementing boundary clock may not have full GRU functions

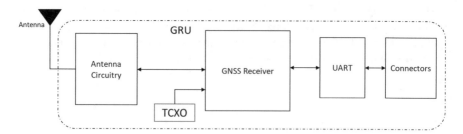

Fig. 5.1 The block diagram of GNSS Receiver Unit (GRU)

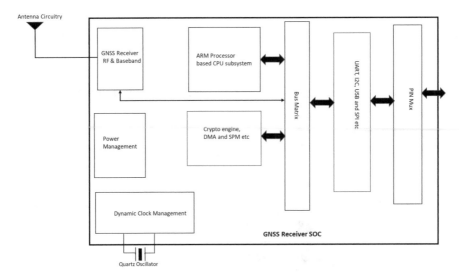

Fig. 5.2 The GNSS receiver SOC block diagram

that are complemented through external GRU. However, newer cell site routers often integrate GRU functions as an embedded timing module to the switch.

5.2.1 Timing Module

An embedded module that integrates much of GRU functions excluding antenna filters and power supply is known as timing modules. This embedded timing modules are widely used and often integrated with networking and industrial equipment. Timing modules are available in various packaging types, e.g., LGA and edge castellation. Typical size of timing module is 19x19mm as shown in Fig. 5.3. The 28 pin Trimble® ICM360 timing module that includes oscillator, GNSS receiver, SAW filter, and various I/O. Trimble ICM360 is often integrated in a network equipment, e.g., DCSG (Disaggregated Cell Site Gateway), Radio Unit, ENodeB, GNodeB, Passive Radar, and PMU (Phasor Measurement Unit).

We will learn about some of these devices in the proceeding chapters. Some of the common timing interfaces available in external GRU and timing modules are 1PPS, 10 MHz, and TOD.

5.2.1.1 1PPS and TOD Interface

ITU-T G.8271 and ITU-T G.703 define 1PPS, TOD, and 10 MHz, respectively. Both 1PPS and TOD use ITU-T V.11 type phase/time distribution interface for time of day (TOD) and one pulse per second (1PPS). The physical connector of V.11

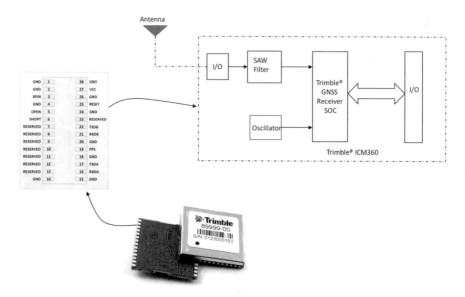

Fig. 5.3 The diagrammatical representation of Trimble® Timing module "ICM 360" (Courtesy: Trimble, Inc)

Table 5.1 1PPS RJ45 connector pin-out for 1PPS input mode as defined in ITU-T G.703

PIN	Signal name	Signal definition
1	Reserved	For further study
2	Reserved	For further study
3	1PPS_IN-	RX 1PPS negative voltage
4	GND	Signal ground
5	User defined	This pin could be used for ground or connection to GNSS receivers. If this pin is not used, ITI-T G.703 recommends to pull it down with a resistor of 10 kΩ
6	1PPS_IN+	RX 1PPS positive voltage
7	RX−	Rx TOD time message negative voltage
8	RX+	Rx TOD time message positive voltage

interface that carries 1PPS and TOD is a RJ45 type [IEC 60603-7] for twisted-pair copper wire medium. Tables 5.1 and 5.2 present RJ45 connector pin-outs for 1PPS and TOD input and output modes as defined in ITU-T G.703.

The 1PPS is a time strobe signal which includes electrical signal with associated time tag messages. The signal is typically 8 µs wide with leading edge of the pulse coincided with the beginning of each UTC second that rise and fall at every 100 ns. It is simplest form of time synchronization and does not contain information about the specific time of day or year. Table 5.3 depicts 1PPS performance parameter as defined in ITU-T G.703. The performance parameters provided in the table below is applicable to the 1PPS interface that is using a 50 Ω 3-meter cable.

Table 5.2 1PPS RJ45 connector pin-out for 1PPS output mode as defined in ITU-T G.703

PIN	Signal name	Signal definition
1	Reserved	For further study
2	Reserved	For further study
3	1PPS_OUT-	TX 1PPS negative voltage
4	GND	Signal ground
5	GND	Signal ground but this pin could be used for ground or connection to GNSS receivers.
6	1PPS_OUT+	TX 1PPS positive voltage
7	TX−	Tx TOD time message negative voltage
8	TX+	Tx TOD time message positive voltage

Table 5.3 1PPS performance parameters as defined in ITU-T G.703

Parameter	Tolerance	Comments
10–90% rise times of the 1PPS	<5 ns	Measured at the 1PPS interface
Pulse width	100 ns to 500 ms	Measured at the 1PPS interface
Maximum cable length	3 m	Due to delay and rise time performance

Although the above table depicts a value of 100 ns to 500 ms for the pulse width, ITU-T G.703 reiterated that the value of 100 ns is proposed, and other values are subject to further study. Most timing systems specify a time message that is transmitted (usually over a serial data port) that gives the time of day (TOD) for each occurrence of the 1PPS signal. The TOD indicates a specific time with hours, minutes, seconds, and the corresponding date.

5.2.1.2   10 MHz Interface

The interface of 10 MHz has many applications as clock input that includes calibration services, scientific instrument, and Passive Bistatic Radar (PBR) system. Figure 5.4 shows a PBR system using 10 MHz to signal generator and its USRP (Universal Software Radio Peripherals). The Passive Bistatic Radar (PBR) that is generally used in air surveillance is a good example of 10 MHz clock input application. Radar was originally developed for military applications such as air defense and surveillance, how it is increasingly used in many civilian applications, e.g., industrial perimeter security, air, terrestrial and marine traffic control, radar astronomy, geological measurements, both small- and large-scale surveillance systems, meteorological predictions, and many more. The PBR system discussed here uses transmitter and receiver that are separated by a distance unlike monostatic radar for which transmitter and receiver are collocated. Its application with SDR (Software Defined Radio) herein USRP is defining future of radar applications in civilian deployments.

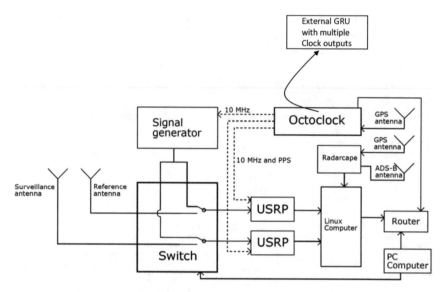

Fig. 5.4 10 MHz clock inputs from GRU with multiple clock outputs in a PBR System (Courtesy: Ali, E. & Östadius, A., 2017) [1]

In the example of PBR deployment, two collocated receiver antennas are used and connected with USRP as shown in the figure above. The Octoclock [1] shown in the figure is a GRU with multiple clock outputs, e.g., 10 MHz and 1 PPS for traceable UTC.

ITU-T G.811 and G.8272 define 10 MHz as output interface for PRC and PRTC including GRU discussed here. T-GM and T-BC may also implement 10 MHz as interface depending upon the product requirements. ITU-T G.703 defines general characteristics of 10 MHz synchronization interface for which signal must confirm to the mask presented in Fig. 5.5.

Table 5.4 shows a general characteristic of 10 MHz synchronization (often referred as sync) interface output signal as defined in ITU-T G.703.

5.2.2 Disciplined Clock

A disciplined clock can be considered a GRU with holdover time capabilities in case of GNSS signal failure. A disciplined timing device is also known as GPSDO (GPS Disciplined Oscillator) and GNSSDO (GNSS Disciplined Oscillator) depending upon if the device can be locked only to GPS or GNSS constellations. The disciplined clock is considered a stable frequency reference device that provides 1PPS and 10 MHz output for frequency reference. It can also provide a holdover from several minutes to 48 h in case of GPS/GNSS signal failure. During the holdover period, it uses a stable oscillator for frequency reference. Figure 5.6 illustrates

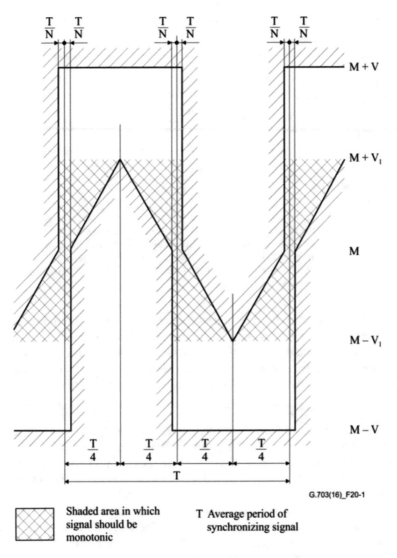

Fig. 5.5 Wave shape of the output of 10 MHz synchronization interface. (Courtesy: ITU-T G.703) [2]

Trimble' Thunderbolt-E disciplined clock that provides up to 24 h holder capabilities. A typical disciplined clock consists of a high-quality (precision) oscillator that is continuously being corrected using the coordinated universal timing signal (UTC), or standard second obtained from the GPS/GNSS constellations.

The satellite signal can be trusted as a reference for two reasons: first, it originates from atomic oscillators, and secondly, it must be accurate in order for GPS/GNSS to meet its specifications as a positioning and navigation system. To understand this further, let's consider that the oscillators onboard a GPS or GNSS satellite

Table 5.4 The general characteristics of 10 MHz synchronization interface

Parameter	Characteristics
Pulse shape	As stated, signal must conform to the mask presented in Fig. 5.5
Test load impedance	50 Ω
Maximum peak voltage (V)	2.5 V
Minimum peak voltage (V1)	0.25 V
Offset voltage (M)	0
Transition region factor (N)	30
Maximum jitter at an output port	The intrinsic jitter at 10 MHz output interfaces as measured over a 60-s interval shall not exceed 0.01 UIpp when measured through a single pole band-pass filter with corner frequencies at 20 Hz and 100 kHz. The jitter unit is called unit interval (UI). A UI is simply a bit period for the considered rate. Peak-to-peak jitter is written UI pp. and a UI pp. gives measure of worst-case jitter

receives clock correction from ground stations once during each orbit (for GPS, it is once every 12 h). The maximum acceptable deviation from the satellite clocks to the positioning uncertainty is generally considered to be about 1 m (meter) for GPS. Since light travels at about 3×10^8 m/s (299,792,458 meters per second) and a meter is defined as the length of the path traveled by light in vacuum during a time interval of $\frac{1}{299792458}$ second or 3.3 ns (approx.), the 1 m requirement is equivalent to a time error of about 3.3 ns. Henceforth, to meet its specifications, a GPS satellite clock must be stable enough to keep time with an uncertainty of less than 3.3 ns during the period between corrections. This translates to a frequency stability specification near 6×10^{-14}. The goal of a disciplined clock is to transfer this inherent accuracy and stability of GPS/GNSS signals to signals generated by OCXO or rubidium oscillator. How to transfer this frequency and time from a master oscillator to local oscillator has been a subject of research for decades. Much of this process of disciplining local oscillator is patented IP (intellectual property) of respective manufacturer; however, there are few basic concepts that apply to those designs.

Generally, the disciplining mechanism works in a similar way to a phase-locked loop (PLL). Some GPSDO/GNSSDOs utilize PLLs; however, many are using microcontroller to compensate for frequency, temperature, and other environmental parameters. A PLL consists of three main elements, a phase detector, a VCO (Voltage Control Oscillator), and a Loop filter. It works by taking signal to which it locks and can then output this signal from its own internal VCO as illustrated in Fig. 5.7.

The phase detector takes reference signal input and signal from VCO; it compares the phase differences between the input signals and produces output signal that is proportional to the difference between the two input signals. This output signal is then passed through the loop filter to reduce high frequency components

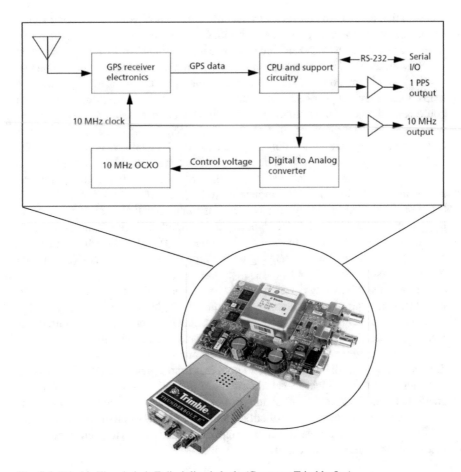

Fig. 5.6 Trimble Thunderbolt-E disciplined clock. (Courtesy: Trimble, Inc)

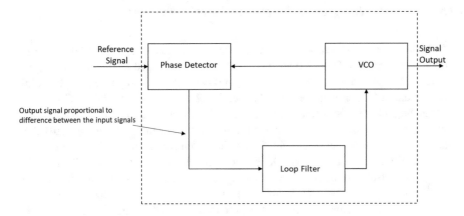

Fig. 5.7 The operation of a Phase-Locked Loop (PLL)

and supplied to VCO to control its frequency. This control voltage as output signal of loop filter changes the frequency of the VCO in a direction that reduces the phase difference between the VCO and the reference input signal. The PLL is locked when the phase of the VCO has a constant offset relative to the phase of the input signal [3].

The disciplining control system similar to that of a PLL described herein above helps correct any jitter, offset, and other deviations in the oscillator; the system could basically count the number of cycles that occurs between the rising edges of two consecutive 1PPS pulses (in 1 s). A typical disciplined clock can track from 8 to 12 satellites and output a 1 pulse per second (PPS) signal synchronized to UTC (USNO). If the disciplined clock or GNSSDO has 10 MHz oscillator and 10,000,010 cycles are counted, then the oscillator is offset of 10. It counts again and new value is 10,000,001 cycles, indicating that the oscillator is now running with 0.1 ppm offset. The control system uses that information to make another correction to perhaps get 9,999,999.9 (−0.01 ppm). The disciplined clock continues this process of finer correction until the count is as close to 10,000,000.00000 cycles as the system can measure, hence achieving a very high frequency accuracy (e.g., in the order of parts-per-trillion, or 1.0E-12) [4].

This process of steering correction also depends on quality of the local oscillator. For example, a rubidium oscillator of high quality might change its frequency due to aging at a rate of less than 1×10–11 per month. However, in case of an inexpensive quartz oscillator it might age 1000 times faster than a rubidium oscillator, so aging compensation will be needed more often and the aging rate will be less predictable [3]. One of the issue here is to note that time signals received from GNSS satellites may not be ideal in certain situation. It may be subject to atmospheric conditions, multipath and other abnormalities that impacts the accuracy and stability of signals. This means that corrections made by disciplined clock may be slightly off. To mitigate this problem, some disciplined clocks use statistical algorithms before issuing the steer correction to the oscillator.

5.2.3   GRU Integrated Antenna (Smart Antenna)

A smart antenna is a variation of GRU that integrates antenna and GRU in one device. An example of such device is Trimble's Acutime 360 as shown in the figure below. The smart antenna typically integrates GRU-type timing module with a high gain antenna. The signals are carried over RS422 cable up to a distance of 4000 feet depending upon the loss calculations. The high gain antenna generally compensates for the loss; however, proper adjustment must be made if loss is more than 5 dB. The smart GNSS antenna shown in Fig. 5.8 uses a 12-pin connector specifically designed withstanding harsh condition and the system adheres to IP67 standard. These devices are useful for applications where space is limited, power outlet is readily not available, and harsh weather condition is also not suitable for external GRU. The radome shape of smart antenna is designed to protect against water and snow

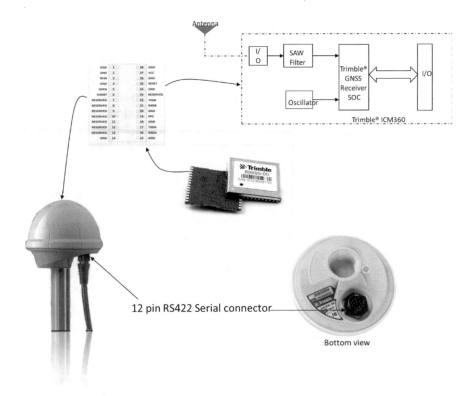

Fig. 5.8   Trimble's Smart Antenna Acutime 360 (Courtesy: Trimble, Inc)

deposits. All signals including UART, 1PPS, power, etc. are carried over a single RS422 cable.

The smart GNSS antenna device acts as single source for RF and PRTC input traceable to UTC with an output of 1PPS and time tag. These antennas can be single band or dual band, GPS only or multi-constellation. Once powered on, a typical smart GNSS antenna such as "Acutime 360" automatically tracks satellites and surveys its position to within meters. It then switches to over-determined time mode and generates a PPS, outputting a time tag for each pulse. The Time-Receiver Autonomous Integrity Monitor (T-RAIM) algorithm of a smart antenna maintains PPS integrity.

The GNSS smart antenna can operate in extreme temperatures (−40 °C to +85 °C) and hostile RF environments typically encountered at wireless network transmitter sites. It requires less than 1 watt of power to operate and outputs the Trimble Standard Interface Protocol (TSIP) or industry-standard NMEA messages [5]. We will discuss about the TSIP, NMEA, and other protocols that are used to control and configure GNSS receiver in the proceeding chapter of the book.

5.3   Packet Timing Clock Devices

A packet network that requires end to end time synchronization generally uses a combination of clocks distribution devices such as grandmaster clock, boundary clock, ordinary clock, and slave clock. These devices are also known as PTP clock devices. The grandmaster clock synchronizes to GPS/GNSS clock using a built-in PRTC while a boundary clock generally acts as slave to grandmaster and provides master clock to slave devices further downstream. The ordinary clock may act as a master or slave clock. In most cases, it remains in slave state until a grandmaster clock fails and no other device capable of providing master clock. A slave clock depends on a boundary clock or grandmaster clock for clock accuracy. A number of standards collectively define grandmaster, boundary, and slave clocks.

Figure 5.9 depicts some of associated standards for ePRTC, grandmaster, boundary, and slave clocks. While grandmaster, boundary clock, and slave are widely used as packet timing devices in the network, ePRTC is a new concept. Defined by ITU-T G.8272.1, an ePRTC is a core packet timing clock that has better performance than a grandmaster and may deploy atomic reference standard for highly accurate time.

5.3.1   Grandmaster Clock

The grandmaster clock or often known as T-GM (Telecom-Grandmaster Clock) is an important timing device in a packet network that combines PRTC function, PTP (Precision Time Protocol) for distribution of phase and time defined by IEEE1588 and switching function in one single device. There are several ITU-T standards which define its functional model and operations. The ITU-T G.8266/Y.1376 standard defines the timing characteristics of T-GM while ITU-T G.8265.1 defines related profile for time distribution by T-GM and the PRTC aspects of the T-GM are defined in ITU-T G.803. Figure 5.10 illustrates typical T-GM functional design. It includes the functional model of a T-GM defined by ITU-T G.8266 and Trimble's Grandmaster clock to depict both functional blocks and physical view of a T-GM device.

The frequency and timescale source clock block synchronizes to frequency and timescale input such as GNSS, 1PPS, and TOD and provides this as a clock source to clock selection block of the T-GM. On the other hand, frequency source block receives frequency input (e.g., SyncE and 10 MHz) and provides this as a reference to clock selection block. The clock selection block selects to local source clock to be used from various inputs and provides this to local clock and timescale generation block. Based upon the input from clock selection block, the local clock and timescale generation block generate local clock and timescale of the clock providing it to the packet processing block.

The packet processing block that works as a PTP (Precision Time Protocol) master clock will insert the local timescale information into the timestamp of PTP

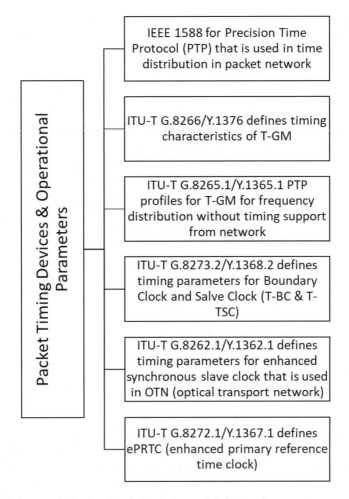

Fig. 5.9 Various standards of packet timing devices and their operations

packet as per the required PTP profile. The processing block also processes ingress and egress PTP packet as per the required profile. We will discuss about PTP and PTP profiles in the proceeding chapter of the book.

The packet I/O is the interface of the T-GM that is responsible for transmission and reception of PTP packets. An ethernet interface is generally considered as packet I/O for the T-GM as shown in the figure above. For frequency accuracy, T-GM should conform to one of three types specified in ITU-T G.812. The following table depicts three different types of node clock that is applicable to T-GM when T-GM is experiencing a prolonged holdover condition. A T-GM may choose to support one such types of frequency output accuracy traceable to a primary reference clock for a time period defined by "Period T" in Table 5.5.

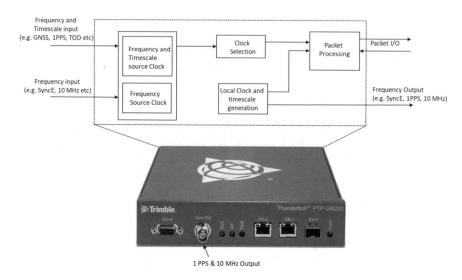

Fig. 5.10 ITU-T G.8266 T-GM functional model and Trimble's Thunderbolt™ GM200 Grandmaster device

Table 5.5 Different types of node clocks defined by ITU-T G.812

Parameter	Type I	Type II	Type III
Accuracy	N/A	1.6×10^{-8}	4.6×10^{-6}
Period	N/A	1 year	1 year

In discussing timing parameters at Chap. 2, PRTC definitions and TE (Time Error) values are discussed. ITU-T G.8272 defines two types of PRTC that are applicable to T-GM: PRTC-A and PRTC-B. A grandmaster clock should implement either PRTC-A or PRTC-B timing criteria to distinguish it. The grandmaster clock (T-GM) that implements PRTC-A should be accurate to within 100 ns or better when verified against primary reference clock traceable to UTC. In this case the value includes time offset and phase wanders of the PRTC-A implementation in a T-GM. The value identified here is for locked operational mode. Similarly, a T-GM implementing PRTC-B should be accurate to within 40 ns or better when compared against primary reference clock traceable to UTC. Time error limits and other values of PRTC-A and PRTC-B as applicable to T-GM are discussed in Chap. 2.

A packet network utilizes grandmaster clock or T-GM to receive UTC-based time reference from GNSS satellites and provide network wide time distribution through PTP timestamps that is traceable to UTC reference in a master slave time synchronization network architecture. The T-GM delivers PTP service with time-stamp traceable primary reference clock over ethernet ports which is then distributed downstream to other clocks such as boundary clock or slave clock. When successfully receiving a reference signal, the grandmaster derives accurate time from the reference. However, when issues occur such as spoofing attacks, grandmasters must be capable of keeping time autonomously, ideally for long periods and with high levels of precision. A grandmaster can implement a range of measures to protect against abnormalities that may occur in a network. These measures include asymmetry compensation for packet network to deal with PDV issues, mechanisms to protect against jamming and spoofing and high-quality crystal oscillator for holdover performance in case of GNSS signal failure.

5.3.2   Boundary and Slave Clock

The boundary clock also known as T-BC (Telecom Boundary Clock) extends the reach of T-GM to further reaches of packet network by acting as a slave for T-GM while providing master clock as a T-GM proxy for the slave devices. The slave clock also known as T-TSC (Telecom Time Slave Clock) may receive master clock from T-BC or directly from T-GM. ITU-T G.8273.2/Y.1368.2 defines timing characteristics for both T-BC and T-TSC. The functional model for T-BC and T-TSC is illustrated in Fig. 5.11.

The packet timing signal is processed by PP block and timestamps are sent to packet time and packet-based equipment clock blocks. The time information present in the timestamp is used to generate the time information to control local timescale. Delay asymmetry may be used as a correction term. The time information present in the timestamp may be used by the PEC to generate local frequency. Time selector block may choose to time information from timestamp or frequency input, e.g., 1PPS + TOD as input. Similarly, frequency selector block may use frequency from timestamp or frequency recovered from physical layer clock as input. The

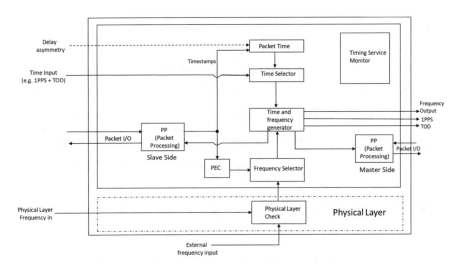

Fig. 5.11 Functional model of T-BC/T-TSC as defined in ITU-T G.8273.2

timing service monitor block monitors PTP timing service according to performance criteria and generates alarms accordingly.

From operational perspective, there are basically two clock inputs for T-BC/T-TSC as shown in Fig. 5.12: a frequency clock locked to the physical layer frequency and a time clock locked to the PTP input. The frequency input and output could be SyncE or SDH. Most commonly, SyncE is used in modern telecom network as the frequency input as it works in tandem with ethernet interface.

The PTP time input is taken to the slave port of the T-BC/TSC and timestamp information in a PTP packet is provided through the master port of T-BC. A slave clock does not provide timestamp input in PTP packet.

It is to be noted that from operational perspective, T-BC and T-TSC must adhere one of the four classes defined in ITU-T G.8273.2 and these are: Class A, B, C, and D. Table 5.6 depicts time accuracy applicable for each class. The maxITEI is one of the most important factors in determining the type of T-BC and sometimes T-TSC can be used for a given deployment.

The values depicted in the table is valid for T-BC/T-TSC with 1PPS and 1GbE, 10GbE, 25GbE, 40GbE and 100GbE. Another important aspect to note is the holdover situation when a failure of clock input occurs. A T-BC must provide holdover capabilities for which there are two types of holdover applicable to both T-BC and T-TSC. First, is the situation when T-BC/T-TSC losses PTP time reference as illustrated in Fig. 5.13.

Let's consider the scenario in which PTP communication failure occurred perhaps due to excessive network delay or jitter between T-BC and T-GM as depicted in the left side of the diagram above. Trimble GM200 can act as T-GM and T-BC, in case of T-BC there is no GNSS input and T-BC is taking master clock input from both PTP and SyncE. Although a T-BC is capable of handling certain level of jitter condition, excessive jitter may cause PTP communication failure. Assume that such

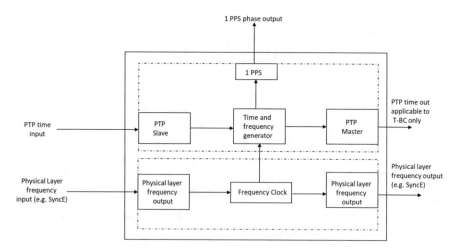

Fig. 5.12 An operational perspective of T-BC/T-TSC as defined in ITU-T G.8273.2

Table   5.6 Maximum absolute time error values for different classes of T-BC/T-TSC

T-BC/T-TSC Class	Maximum absolute time error max|TE| (ns)
A	100 ns
B	70 ns
C	30 ns
D	Subject to further study

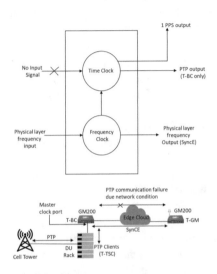 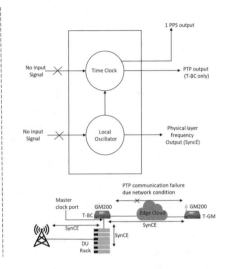

Fig. 5.13 T-BC/T-TSC holdover situation

condition occurred momentarily in the network, a T-BC will fall back to SyncE and provide master clock to PTP clients that are acting as T-TSC. In this case a 5G deployment scenario is considered having a DU (Distributed Unit) rack that are connected to radio units (RU) at cell towers. Both DUs and RUs are capable of implementing T-TSCs with PTP software client or PTP client with hardware time-stamping capabilities. They also implement SyncE capabilities as well and thus receive master clock from T-BC as PTP input and/or SyncE input. When the PTP communication failed between T-GM and T-BC, the T-BC will use SyncE to retrieve master clock and in turn provide master clock output to PTP packets and SyncE for downstream clients.

Now, let's assume that PTP communication failure occurred throughout the network (please refer to right side of the diagram). In this case, RU and DUs are also not receiving master clock from PTP. They fallback to SyncE and retrieve the master clock from SyncE for which a T-BC is providing master clock in the form of physical layer frequency output.

5.3.3   Enhanced Primary Reference Time Clock (ePRTC)

The ePRTC is a new addition to packet network timing device. It provides atomic frequency standard capabilities and acts as core clock for the network or a section of network. ITU-T recommendation G.8272.1/Y.1367.1 defines the timing characteristics of ePRTC. Compared to T-GM with PRTC-A and PRTC-B, e PRTC is subject to more stringent requirements and includes a frequency input directly from an autonomous primary reference clock (e.g., atomic clock). The wander on a frequency interface from the autonomous primary reference clock (e.g., atomic clock) expressed in TDEV should follow the limits specified in Table 5.7.

Please note GNSS receiver clock input is not considered an autonomous primary reference clock for ePRTC and for this reason an ePRTC must have atomic clock implemented through cesium or rubidium crystal. As the core clock, an ePRTC provides reference signal for time, phase, and frequency for clocks within a network or a section of network including T-GM.

Figure 5.14 depicts the functional model of ePRTC. The output shown is logical interface of ePRTC. In some implementation, various logical interfaces can be combined in a physical interface.

The ePRTC must implement an external input, e.g., given cesium input as autonomous primary reference clock as shown in the figure and implement at least one frequency output interface. Finally, an ePRTC may also deliver traceability information, reflecting the status of the clock (i.e., locked on its input reference signal, in holdover). It is to be noted that traceability information is subject to further study.

The time recovery block receives traceability information through GNSS receiver/GNSS antenna and generates frequency, phase, and time output. Local frequency clock block synchronizes with cesium input and provides a reference to local timescale block which maintains the local representation of the primary timescale based on the input from local frequency clock block. It then generates time and phase reference as output.

When an ePRTC losses all its input and enters to holdover state, it must rely on autonomous primary reference clock (e.g., atomic clock) input. ITU-T G.8272.1 defines two ePRTC classes that are subject to stringent holdover requirements: ePRTC-A and ePRTC-B. For ePRTC-A, the holdover period is 14 days and ePRTC-A should be accurate, when verified against the applicable primary time standard (e.g., UTC), to within a value increasing linearly from 30 to 100 ns [6].

Table 5.8 depicts phase/time holdover for ePRTC-A, whereas the parameters for ePRTC-B are subject to further study.

During locked mode as in normal operation, if the reference time signal from GNSS is subject to multipath reflections and interference from other local

Table 5.7 Wander Generation (TDEV) for autonomous primary reference clock of ePRTC as defined in ITU-T G.8272.1

TDEV limit (ns)	Observation Interval τ (s)
1	$0.1 < \tau \le 10{,}000$

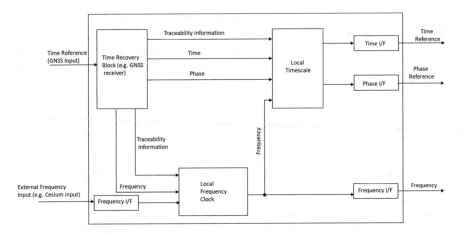

Fig. 5.14 Functional model of ePRTC as defined by ITU-T G.8272.1

Table 5.8 ePRTC Phase/time holdover requirements as defined in ITU-T G.8272.1

ePRTC class	Time t (s)	Time error $\Delta x(t)$ (ns)
ePRTC-A	$0 < t \le 1{,}209{,}600$ (14 days)	$\lvert \Delta x(t) \rvert \le 30 + 5.787037 \times 10^{-5}\, t$
ePRTC-B	Subject to further study	Subject to further study

Note: $t = 0$ refers to start of holdover. A longer holdover duration around 80 days with <100 ns accuracy is under discussion for ePRTC-B

Table 5.9 Wander Generation (MTIE) of an ePRTC

MTIE limit (ns)	Observation interval τ (s)
4	$0.1 < \tau \le 1$
$0.11114 \times \tau + 3.89$	$1 < \tau \le 100$
$0.0375 \times 10^{-3}\, \tau + 15$	$100 < \tau \le 400{,}000$
30	$\tau > 400{,}000$

Table 5.10 Wander generation measured in TDEV

TDEV Limit (ns)	Observation interval τ (s)
1	$0.1 < \tau \le 30{,}000$
$3.33333 \times 10^{-5}\, \tau$	$30{,}000 < \tau \le 300{,}000$
10	$300{,}000 < \tau < 1{,}000{,}000$

transmissions, such as jamming, an ePRTC must be minimized to an acceptable level. The wander measurements for this normal state of operation are depicted in Tables 5.9 and 5.10. The parameters include both MTIE, a measurement related to peak-to-peak wander and TDEV, a measurement related to rms wander.

Additionally, ITU-T G.8272.1 specifies that an ePRTC to be accurate to within <30 ns when verified against the applicable primary time standard. The requirement of 30 ns is driven by the need to maintain small overall error budget as time is passed from the core to the edge of the network. Henceforth, if an ePRTC has a greater accuracy it will help maintain small error budget downstream in the network. From this perspective and considering its longer holdover capabilities, an ePRTC provides stable and dependable clock services for the network while helping with smaller overall error budget to keep time and phase alignment optimized throughout the network.

References

1. Ali, E., & Östadius, A. (2017). *Passive radar detection of aerial targets*. Master thesis, Lund University.
2. G.703. *Physical/electrical characteristics of hierarchical digital interfaces*. ITU-T G.703 (04/2016). International Telecommunication Union.
3. Lombardi, A. M. (2008). *The use of GPS disciplined oscillators as primary frequency standards for calibration and metrology laboratories*. Technical Paper. National Institute of Standards and Technology.
4. Polo, M. I. (2016). *GPS disciplining and holdover for field testing introduction, examples, analysis and recommendations*. VeEX Inc.
5. GPS world. (2016). *Trimble multi-GNSS timing antenna allows for BeiDou, Galileo*. GPS World. Available online at https://www.gpsworld.com/trimble-multi-gnss-timing-antenna-allows-for-beidou-galileo/
6. G.8272.1. (2016). *Timing characteristics of enhanced primary reference time clocks*. ITU-T G.8272.1/Y.1367.1 (11/2016). International Telecommunication Union.

Chapter 6
Method of Time Distribution

6.1 Introduction

The chapter begins with discussion of GRU interface and configuration protocols providing readership a brief overview of how GRU is configured and communication parameters are retrieved and processed between GRU and user equipment. This understanding is important since GRU in its various forms provide primary reference clock traceable to UTC and functional elements of GRU are often implemented in packet timing device such as T-GM. Later, various timing distribution methods are introduced which can be divided into two categories: frequency-based timing distribution and packet-based timing distribution. The chapter refers to various methods for the both type of distributions. However, frequency-based timing distribution is further elaborated and various methods such as BITS/SSU, IRIG, IRIG-B, and SyncE (Synchronous Ethernet) are introduced. Though BITS/SSU may be concern case where telecom and CATV service providers have some install base, IRIG-B and SyncE are common. Many power utilities still use IRIG-B timing codes for their network in a hybrid mode with some SyncE and packet timing protocols. Overall, SyncE is gaining momentum in terms of frequency-based timing distribution and works in tandem with packet timing protocols for asynchronous networks due to increased penetration of ethernet. This chapter provides the insights of operational and functional models of BITS/SSU, IRG-B, and SyncE. Readerships will find these insights useful as they design network synchronous with backward and forward compatibility.

© The Author(s), under exclusive license to Springer Nature
Switzerland AG 2021
D. D. Chowdhury, *NextGen Network Synchronization*,
https://doi.org/10.1007/978-3-030-71179-5_6

6.2   GNSS Interface and Configuration Protocols

In Chap. 5, we discussed and learned about the importance of GRU as a timing device. Most GRUs with its variants in the form of external GRU, GNSS receiver timing modules, disciplined clocks, and smart antennas use one of the four protocols to control and configure GRU or GNSS receiver units: TSIP (Trimble Standard Interface Protocol), NMEA (National Marine Electronics Association), and RTCM (Radio Technical Commission for Maritime). From operational perspectives, these protocols are essential to set up GNSS receivers and steer them appropriately to control beacon and satellite parameters.

6.2.1   Trimble Standard Interface Protocol (TSIP)

Developed by Trimble, Inc., Trimble Standard Interface Protocol (TSIP) provides the system designer with over 75 commands that help configure a GPS/GNSS receiver for optimum performance in a variety of applications. This protocol was originally defined for the Trimble Advanced Navigation Sensor (TANS) and referred to as TANS protocol even though it was applied to many other devices.

Typically, one serial port of a GPS/GNSS receiver unit is used for both input and out TSIP commands and reports. TSIP is based on the transmission of packets of information between the user equipment and the GPS/GNSS unit. Each packet includes an identification code (1 byte, representing 2 hexadecimal digits) that identifies the meaning and format of the data that follows. Each packet begins and ends with control characters. Figure 6.1 illustrates the packet format of TSIP protocol.

The <DLE> indicates start of packets, one byte packet identifier <ID> can be any value excepting <DLE> and <ETX>. The bytes in the data bytes or data string bytes field can have any value. To prevent confusion with the frame sequences <DLE> <id> and <DLE> <ETX>, every <DLE> byte in the data string is preceded by an extra <DLE> byte ("stuffing"). These extra <DLE> bytes must be added ("stuffed") before sending a packet and removed after receiving the packet. It is to be noted that a simple <DLE> <ETX> sequence does not necessarily signify the end of the packet, as these can be bytes in the middle of a data string. The end of a packet is <ETX> preceded by an odd number of <DLE> bytes.

6.2.2   NMEA Protocol

Defined by National Marine Electronics Association (NMEA), NMEA protocol, "NMEA 0183" combines electrical and data specification for communication between GNSS receivers and user equipment. It replaces the earlier NMEA 0180 and NMEA 0182 standards. NMEA 0183 uses a simple ASCII serial

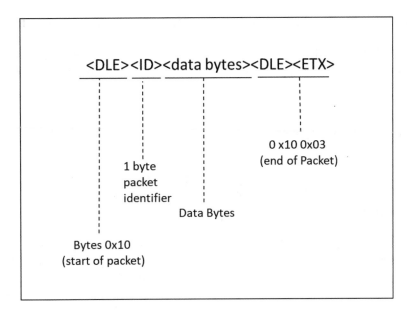

Fig. 6.1 TSIP Packet Structure

Table 6.1 NMEA 0183 serial communication

Signal	NMEA standard
Baud rate	115 kbps
Data bits	8
Parity	None (disabled)
Stop bits	1

communications protocol that defines how data can be transmitted in a "sentence" from one "talker" to multiple "listeners" at a time. NMEA 0183 specification recommends that that the talker output comply with EIA RS-422. The protocol allows a single source (talker) to transmit serial data over a single twisted wire pair to one or more receivers (listeners). Table 6.1 lists the standard characteristics of the NMEA 0183 data transmissions.

All data transmitted over serial communications must be in the form of sentences. Only printable ASCII characters are allowed, plus CR (carriage return) and LF (line feed). Each sentence starts with a "$" sign and ends with <CR><LF>.

An example of NMEA 0183 message structure is as follows:

$IDMSG,D1,D2,D3,D4,...,Dn*CS[CR][LF]

The "$" signifies start of sentence message followed by "ID," a two-letter mnemonic that describes the source of the navigation information. For example, GP identifies GPS and GL identifies GLONASS source. The "ID" follows by a three-letter mnemonic "MSG" that describes message content and the number and order of the data fields. The "," after "MSG" serves as delimiter for data field. Each message contains multiple data fields (D1, D2, D3, D4,... Dn) each of which is delimited by commas ",". The asterisk "*" serves as checksum delimiter and "CS" is the checksum field that contains two ASCII characters describing hexadecimal value of checksum. The [CR][LF] is the carriage return and line feed respectively to terminate the message.

6.2.3   RTCM Protocol

The RTCM (Radio Technical Commission for Maritime) is an international non-profit scientific, professional and educational organization that came up with a way to communicate positions for boats and other vessels many decades ago. This protocol was originally developed by RTCM Special Committee (SC-104) in 1985 as open standard protocol that is used to supply the GNSS receiver with real-time differential correction data. It came a long way since its inception. For example, RTCM 2.0 supported only GPS code which augmented in RTCM 2.2 that supports GLONASS and version 2.3 included antenna corrections, and the changes continued. In 2007, the RTCM Special Committee 104 published RTCM 3.0 that utilizes a more efficient message structure than its predecessors supporting RTK (Real-Time Kinematic) communications. The RTK is a technique used to increase the accuracy of GPS signals by using a fixed base station that wirelessly sends out corrections to a moving receiver. RTCM version 3.0 also provides both GPS and GLONASS code and carrier messages, antenna and system parameters. RTCM 3.1 adds a network correction message and version 3.2 announced in 2013 introduced a feature known as Multiple Signal Messages (MSM) which includes the capability to handle the European Galileo and the Chinese Beidou GNSS systems.

6.3   Time Distribution

The job of a PRTC (whether GRU or integrated PRTC in a T-GM) is to continually sync with Universal Coordinated Time (UTC) while providing input to T-GM and other timing devices for distribution of UTC time across the networks. It may use a combination of GPS, GNSS, and other timing input mechanisms (e.g., atomic clock in ePRTC) to achieve this. There are several mechanisms to carry timing information throughout the networks that includes BITS (Building Integrated Timing Supply), IRIG (Inter-Range Instrumentation Group) time code type B (IRIG-B), 1PPS (1 Pulse per Second), and SyncE. All these technologies are dedicated timing signals requiring a physical connection specifically for timing. One exception is that SyncE can coexist in a physical connection of a packetized network; in other words, same port of an ethernet switch can implement SyncE and transport packets for shared physical link.

Apart from dedicated timing signals, there are other packet base solutions for timing distribution as well, e.g., NTP (Network Transport Protocol) and PTP

(Precision Time Protocol). Both protocols require no specific connection for timing and are best suited for packetized networks. While NTP is a common time distribution protocol for computer networks and in existence for nearly three decades, PTP (IEEE 1588) is relatively new. It is defined by IEEE 1588 specification in 2002. Since its inception, PTP has gained increased attention due to the possibility of achieving sub-nanosecond accuracy when used in conjunction with PRTC for primary clock source and SyncE to distribute timing information.

Table 6.2 shows different methods of timing distribution and relative timing accuracy for each. A point to note here is that both NTP and PTP support TOD, phase, and frequency synchronization making them ideal for today's packetized networks. However, NTP is less suitable for applications and network where sub-nanoseconds to microsecond accuracy is needed, e.g., 5G TDD deployments such as CBRS, mmWave, etc.

All time distribution methods should adhere to respective standards, e.g., ANSI, Telcordia and ITU-T requirements for PRC (Primary Clock Source) or PRS (Primary Clock source) and time synchronization mechanisms. In a typical deployment, Stratum 1 level clock is considered as PRC/PRS for the network. A Stratum 1 is part of clock hierarchy level defined by ANSI for which Stratum 0 is atomic clock that provides input to Stratum 1. Where atomic clock inputs are not available, the PRC/PRS may take input from GPS/GNSS or cesium clock and a combination thereof as required. At Stratum 2 level, time servers generally get time reference from Stratum 1. The sync plane design should consider respective ANSI and ITU-T standards together for optimal outcomes. It is also useful to define a sync plane that is backward compatible. The figure below shows a relative map between ANSI clock hierarchy and ITU-T recommendations for frequency plane and time/phase plane (e.g., ITU-T PTP profile).

This relative map as presented in Fig. 6.2 is not an exact representation rather an attempt to broaden the understanding of clock hierarchy levels for timing source and distribution. Such understanding will help readership to implement the concept

Table 6.2 Methods of timing distribution

Method of timing distribution	Time of day (TOD)	Phase	Frequency	Accuracy	Topology
PTP (IEEE 1588)	Yes	Yes	Yes	Sub ns to 100 μs	LAN/WAN
NTP	Yes	Yes	Yes	100 μs to ms	LAN/WAN
SyncE	No	No	Yes	Sub ns [note]	LAN
IRIG-B	Yes	Yes	Yes	10 μs to sub ms	Point to point
PPS	No	Yes	Yes	<100 ns	Point to point
BITS	No	No	Yes	<100 ns	Point to point

Note: A network with SyncE (Synchronous Ethernet) and PTP together can achieve sub ns accuracy as evident in white rabbit experiment by CERN

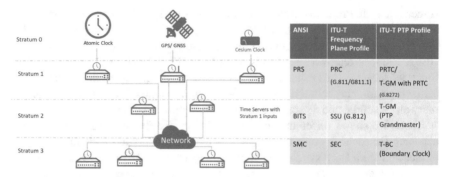

ANSI	ITU-T Frequency Plane Profile	ITU-T PTP Profile
PRS	PRC (G.811/G811.1)	PRTC/ T-GM with PRTC (G.8272)
BITS	SSU (G.812)	T-GM (PTP Grandmaster)
SMC	SEC	T-BC (Boundary Clock)

Fig. 6.2 Clock hierarchy levels for timing source

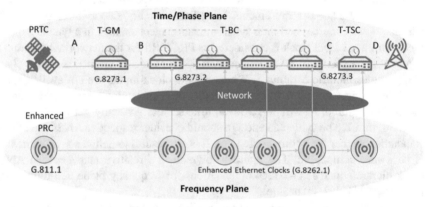

Fig. 6.3 Timing distribution and applicable ITU-T standards in frequency and time/phase plane [1]

in network synchronization design. For a given network synchronization deployments, e.g., fronthaul, ITU-T recommendations should be understood in two distinct planes: frequency and time/phase. In the frequency plane, a set of ITU-T recommendations defines characteristics of the clock and timing distribution: G.811 and G.811.1 define PRC and enhanced PRC (ePRC), respectively (Fig. 6.3).

SyncE (Synchronous Ethernet) is a good example of frequency plane timing distribution. Similarly, PTP (a protocol set defined by IEEE 1588) timing distribution can be better understood by applying time/phase plane characteristics and requirements set forth in ITU-T recommendations G.8271, G.8272, etc. as depicted in the diagram above. Please note, frequency and time/phase sync planes can be managed and routed independently.

6.3.1   BITS and SSU

The BITS (Building Integrated Timing Supply) is a North American term that describes a building-centric timing system. It is also known as SSU (Synchronization Supply Unit) in Europe and other countries. BITS and SSU clock distribution methods are based on a hierarchical structure in which timing is passed down from a master timing source, a Stratum I or PRC clock, to slave clocks (BITS/SSU). A slave clock can pass timing information down to other slave clocks given that the highest performance clocks are at the top of the clock distribution network. Both BITS and SSU master clock distribution methods are considered older and generally used in T1/E1 (TDM) and SONET (Synchronous Optical Network)/SDH (Synchronous Digital Hierarchy) network environment. It is less likely that readership will encounter TDM/SONET/SDH network; however, there are some remnants of these networks still existing in telecom environment and hence, learning of these timing distributions technologies will be useful. The BITS clock receiver will extract timing T1, fractional T1, T3 and T4 signal information from a dedicated timing source. This timing signal then becomes the master timing reference for a given network element. Figure 6.4 shows a typical TDM network with BITS timing distribution. BITS and SSU based timing distribution are mainly a frequency distribution

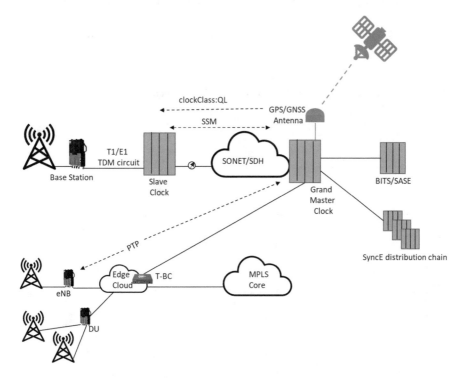

Fig. 6.4 Typical TDM/SONET network with BITS timing distribution

as downstream nodes obtain clock information through frequency synchronization with upstream nodes.

The incoming T1 (DS1) signal can carry traffic or just timing information; but, by using a dedicated timing line, performance, maintenance, and reliability of the BITS clock can be increased. The SSU works similar ways as the BITS clock except that E1 timing levels are used. The signal that is selected for the primary clocking source of the BITS should provide the best level of clock available. Line length, downtime, and jitter are line characteristics that should be considered when selecting the primary clock.

Many traditional CATV and telecom providers still have remnant of SONET/SDH network that requires integration with timing distribution mechanism used in ethernet-based packet network. As these providers are migrating from traditional SONET/SDH to modern ethernet-based transport system, a universal timing distribution is needed that accommodates the need for both networks. Many PTP T-GM and T-TSC today support frequency distribution. This is possible due to frequency distribution profile defined by ITU-T G.8265.1 and known as telecom profile. It allows among other thing backward compatibility with SONET/SDH network. In order to provide UTC traceability in the same way for existing SONET/SDH and SyncE-based synchronization system, the QL indicators defined in G.781 must be carried over the PTP path. ITU-T G.8265.1 defines mapping of SSM (synchronization status messaging) QL (quality level) to the PTP clockClass attributes as shown in Table 6.3.

The three options for SSM QL messages shown in the table are defined in ITU-T G.781.

Table 6.3 Mapping of SSM QL to PTP clockClass values as defined in ITU-T G.8265.1

SSM QL	ITU-T G.781			PTP clockClass
	Option I	Option II	Option III	
0001		QL-PRS		80
0000		QL-STU	QL-UNK	82
0010	QL-PRC			84
0111		QL-ST2		86
0011				88
0100	QL-SSU-A	QL-TNC		90
0101				92
0110				94
1000	QL-SSU-B			96
1001				98
1101		QL-ST3E		100
1010		QL-ST3/ QL-EEC2		102
1011	QL-SEC/ QL-EEC1		QL-SEC	104
1100		QL-SMC		106
1110		QL-PROV		108
1111	QL-DNU	QL-DUS		110

These QL indicators are carried in the SSM message field of Synchronous Optical Network (SONET)/SDH systems and the Ethernet synchronization message channel (ESMC) messages of SyncE where SyncE is used in a network. Similar information is carried in the clockClass field of the PTP announce messages, and various ranges of clockClass values have been designated for use by alternate profiles as shown in the table above. The profile encodes of the QL indicators in the clockClass values 80 to 110.

The SSM message discussed above is defined in ANSI (T1.105) and ITU-T (G.703) recommendations that identifies the quality level of the incoming clock. ITU-T recommendation G. 8264 defines SSM and ESMC for protocol behavior and format for frequency transfer and timing distribution in a network. For SONET/SDH, the SSM is a message contained in the multiplex layer overhead and carries an information about the quality level of the source clock from clock to clock along the branches of the synchronization distribution tree. The source clock can be either the PRC (normal condition) or another clock in holdover mode (failure condition). There are a number of predefined QLs as shown in Table 6.1 corresponding to existing clock specifications, i.e., QL-PRC, QL-SSU-A, QL-SSU-B, QL-SEC, and QL-DNU. In SDH and SONET, SSM is contained in the SSM Byte (SSMB) of the STM-n or OC-n frame overhead. Figure 6.5 shows SSM communications between two network elements (NE) in a SONET/SDH network.

Logically, the SSM overhead can be viewed as a dedicated unidirectional communication channel between entities (herein two NEs as shown in the figure above) that process SSM messages. The simplified example presented in the figure above includes two NEs each connected to a SSU and to each other. The selectors within each NE facilitate source selection for system clock and operates under the control of "Sync Control" block. The "sync control" block is also responsible for controlling timing protection and may have an interface to the management system. This block may take QL SSM as input and may also be responsible for generating an SSM on the appropriate outputs to indicate certain conditions, e.g., insertion of DNU (do not use) on some ports. The DNU (QL-DNU) as depicted in Table 6.3 indicates that the signal carrying this SSM shall not be used for synchronization

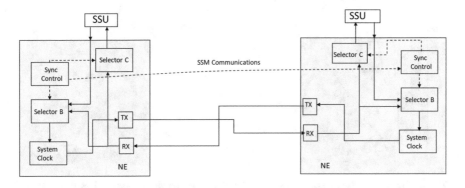

Fig. 6.5 SSM communications between two NEs in a SONET/SDH network [2]

because a timing loop situation could result if it is used. The SSM unidirectional communication between two "sync control" blocks of NEs provides an indication of QL in the transmitting clock.

Typically, a NE has two reference clocks and generates their own synchronization signal in case it loses the external reference. If such is the case, it is said that the NE is in holdover. A synchronization signal that passes through SONET/SDH network must be filtered and regenerated by all the nodes that receive it, since it degrades when it passes through the transmission path. For SONET/SDH, the synchronization network is partially mixed since some NEs may transmit data and distribute clock signals to other NEs. The most common topologies for synchronization in a SONET/SDH networks are: tree, ring, distributed, and meshed topologies. The ring topology relies on a master clock whose reference is distributed to the rest of the slave clocks. It has two weak points: it depends on only one clock, and the signals gradually degrade as depicted in Fig. 6.6.

For SDH networks, different types of SSU are used along with SEC (Synchronous Equipment Clock). Timing characteristics for each are defined in G.812 and G.813, respectively. Table 6.4 depicts different clock accuracy level including those for SSU-T (SSU Transit) and SSU-L (SSU Local) for SDH network.

To transport a clock reference across SDH/SONET, a line signal is used instead of the tributaries transported. The clock derived from a STM-n/OC-m interface is only affected by wander due to temperature and environmental reason. However, care must be taken with the number of NEs to be chained together, as all the NEs regenerate the STM-n/OC-m signal with their own clock and, even if they were well synchronized, they would still cause small, accumulative phase errors.

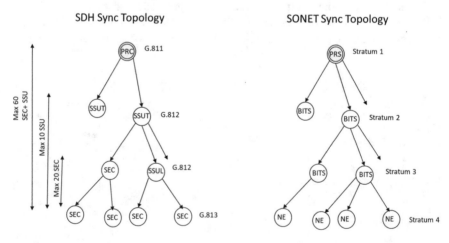

Fig. 6.6 Tree topology for SDH and SONET network synchronization using SSU/SEC and BITS, respectively

Table 6.4 Various clock source for SDH network synchronization

Clock source	Accuracy	Drift	ITU-T standard
PRC	1×10^{-11}	–	G.811
SSU-T	5×10^{-10}	10×10^{-10}/day	G.812
SSU-L	5×10^{-8}	3×10^{-7}/day	G.812
SEC	4.6×10^{-6}	5×10^{-7}/day	G.813

Table 6.5 IRIG time code format [3]

IRIG time code format	Pulse rate (or bit rate)	Index count interval
IRIG-A	1000 PPS (pulse per second)	1 ms
IRIG-B	100 PPS	10 ms
IRIG-D	1 PPM	1 min
IRIG-E	10 PPS	100 ms
IRIG-G	10,000 PPS	0.1 ms
IRIG-H	1 PPS	1 s

6.3.2   IRIG-B Time Code

The IRIG time codes were originally developed by the Inter-Range Instrumentation Group (IRIG), a part of the Range Commanders Council (RCC) of the US Army. The standard was first published in 1960 and described in IRIG Document 104-60. Since then several revisions (IRIG Standard 200-98 and RIG Standard 200-95) of IRIG time codes have published by the Telecommunications and Timing Group (TTG) of the RCC. The latest version is IRIG standard 200-04, "IRIG Serial Time Code Formats," updated in September, 2004. This protocol can be propagated through different media, e.g., coax cable, symmetrical twisted-pair cable, or fiber optic cable. IRIG DC (Digital Current) signal is generally transmitted over coaxial or RS485 or fiber optic cable, whereas IRIG-AM (Amplitude Modulation) is provided on coaxial cable.

The IRIG standard defines a family of six rate-scaled serial time codes; however, IRIG-B is well-known among them. Table 6.5 shows different IRIG time codes.

The IRIG time codes presented in the table above have different time frame and index count. These time codes are transmitted in form of pulse which is identified as single "bit" and the repetition rate is identified as bit rate. The time interval between the leading edge of two consecutive bits is the index count interval and the time frame is duration in which complete IRIG frame is transmitted. For example, IRIG-A signal has time frame of 0.1 s with index count of 1 ms. It contains time information of year in Binary Code Decimal (BCD) format and seconds of day information in Straight Binary Seconds (SBS) format. IRIG-B time code has a time frame of 1 s with an index count of 10 ms and contains TOY (Time of Year) in days, hours, minutes, seconds, and year information in a BCD format and seconds of day in SBS. IRIG-D Time code has a time frame of 1 h with an index count of 1 min and contains TOY information in days and hours in a BCD format. IRIG-E has time frame of 10 s with index count of 100 ms. IRIG-G has a time frame of 0.01 s with an index count of 0.1 ms and contains TOY information in days, hours, minutes, seconds, tenths, and hundredths of seconds and year information in a BCD format. IRIG-H has time frame of 1 min with index count of 1 s.

All time frame begins with reference position market bit P0 and then with reference identifier bit Pr (please refer to Fig. 6.7). The beginning of each 1 s time frame is identified by two consecutive 8 ms bits, P0 and Pr.

The leading edge of Pr is the on-time reference point for the succeeding time code words. Position identifiers, P0 and P1 through P9 each use 10 ms of the time frame, one full index count duration. Time code bit rate for the frame is 100 pps. The three time code words and the CFs presented during the time frame are pulse-width coded [4].

For IRIG-B time code, each pulse is 10 ms duration for which reference and position identifier is of 8 ms duration, Bit 1 is of 5 ms duration and Bit 0 is of 2 ms duration of index count interval as shown in Fig. 6.8.

IRIG-B frame has multiple frame formats: IRIGB120 and IRIG-B122. The IRIGB120 is pulse-width format and IRIG-B122 is amplitude modulated signal. Both of these IRIG-B frame types have TOY information in BCD format. IRIG-B120/122 frame begins with position identifier and reference identifier bits followed by 30 bits consisting of TOY information including seconds, minutes, hours, and days of year.

The IRIG B12x time code signal is generated using microcontroller as depicted in Fig. 6.9. The 1PPS signal from GRU (GPS/GNSS Receiver Unit) is provided as input to microcontroller which uses a timer of 1 ms which is calibrated at every 1PPS signal as such that 1PPS is equal to 1000 ms counts of timer.

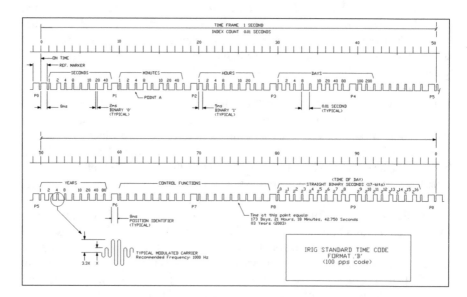

Fig. 6.7 IRIG-B frame format (Courtesy IRIG, 2016) [4]

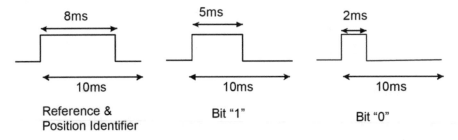

Fig. 6.8 IRIG-B bit format (Courtesy Rajput, 2014) [5]

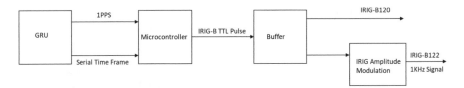

Fig. 6.9 IRIG-B frame generation [5]

This calibration happens continuously to ensure that 100pps of IRIG-B signal is transmitted at every rising edge of 1PPS signal. After the timer of 1 ms is calibrated, the other timer is started which can provide pulse-width modulation (PWM) output of total period 1 ms pulse with minimum duty cycle of 20% and maximum duty cycle of 80%. This PWM-based internal time is also calibrated at every 1PPS signal [5].

After both timers are calibrated, time information is decoded and inserted in IRIG frame at the microcontroller. This data is then transmitted at every 1 ms PWM timer based on 1PPS signal. For distribution of time, IRIG signals whether modulated or demodulated are distributed electrically through simple twister pair or coaxial cable. A typical deployment of IRG-B is power substation where electrical noise limits the max distance for a single segment cable run up to 100 ft. and in some cases 50 ft. Longer distance is achievable by converting ground-referenced signal to a differential signal similar to RS422. Nonetheless, distribution of IRIG-B signal through coaxial or twister pair cable must be limited to the same building. Extending cable to outside building or the electrical apparatus in a longer distance should be done through fiber optic cable. It is to be noted that multiplexing IRIG-B signal with other data over a fiber optic cable may create latency issues which may vary depending on the modulation technique and the relative priority given to IRIG-B data in comparison to other data.

Figure 6.10 illustrates typical IRIG setup in a power station. GPS clock is taken as input for IRIG and passed over RG/58 coax cable to communication processors which then distributes to various IEDs (Intelligent Electronic Devices) such as circuit breakers, transformers, and relays. Use of a communication processor may contribute to latency issues; however, it provides the capability to convert from coaxial to fiber optics connectivity for IEDs that are placed outside the building in relatively

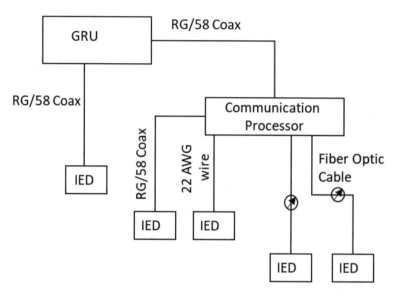

Fig. 6.10 Typical IRIG time code distribution in power substation

longer distance. A typical coaxial and twisted-pair cable up to 135 ft. in length may add 0.25 μs delay. We will discuss further on power substation timing distribution including IRIG-B and PTP interconnectivity in the proceeding chapter.

6.3.3 Synchronous Ethernet (SyncE)

While discussing about BITS and SSU, we explored the potential install base of TDM circuitry including T1/E1 and SONET/SDH. Some of the traditional CATV and telecom providers still have TDM install base which they prefer to integrate with prevailing ethernet technologies. Key to this migration is the network synchronization that for TDM circuitry was mainly based on frequency and node to node interconnectivity for clock transfer. ITU-T G.8261 describes various level of synchronization methods across the network including packet-based methods (e.g., NTP and PTP) and synchronous ethernet (SyncE). For SyncE, ethernet physical layer is used to transport synchronization. The clock signal can be generated from "bit stream" in a similar manner as it is generated in traditional SONET/SDH/PDH networks. Each node in the SyncE chain recovers the clock from the upstream node and distributes the clock to the downstream node by relying on the ethernet physical layer for the transport of the clock. Since it uses the physical layer of ethernet to recover the timing, it can only distribute frequency and not ToD or phase alignment. The concept for ethernet physical layer transport of ethernet equipment clock (EEC) is described in ITU-G.8261 and the mechanism allows support for timing distribution in a manner consistent with SDH frequency-based timing distribution. It is to be noted that the performance of EEC for a SyncE timing distribution is independent of network loading and not influenced by impairments associated with packets networks, e.g., queuing, routing, and PDV. Figure 6.11 shows the number (N) of EECs that can be connected in a series for SyncE timing distribution: usually it is less than or equal to 20 ($N \leq 20$).

The clock synchronization SyncE network is managed in hierarchy similar to SONET/SDH with PRC as the first level and SSU/BITS is the second level in the hierarchy. The second level of the hierarchy herein BITS/SSU requires a holdover feature that allows the unit to maintain timing for some period in case there is a loss of the PRC input in the network. The third level is the EECs connected in series for which the total EECs in series is <20. Similar to SSU, EECs also require holdover, but the accuracy of the holdover is less stringent than the BITS/SSU level. The EECs are implemented with a phase-locked loop (PLL) capable of tracking a local backup OCXO or TCXO reference with frequency stability of ± 4.6 ppm or better. Today's mobile networks require an accuracy of ± 50 ppb. This is achievable in a SyncE network if EEC is fixed to traceable PRC and does not lose connectivity. However, in case of a failure EEC needs local oscillator such as OCXO/TCXO for reference.

As illustrated in Fig. 6.12, a native ethernet network comprising connected ethernet switches has their physical layer (EPHY) accuracy at ± 100 ppm and no traceable primary reference clock. However, a SyncE ethernet network has connected ethernet switches each having a PLL connected to EPHY with traceable reference clock and their accuracy is at ± 4.6 ppm. However, it is to be noted that new specification of EEC (defined by ITU-T G.8262.1) or better known as eEEC (Enhanced Ethernet Equipment Clock) is about five times more stringent than the existing EEC specification (G.8262). The following figure depicts wander generation (MTIE/TDEV) for EEC defined by ITU-T G.8262 and eEEC defined by ITU-T G.8262.1. Applying the MITE mask, the difference can be better understood. An MTIE of 40 ns (the existing G.8262 specification for observation intervals up to 1 s) should require a test instrument with an accuracy of 4 ns or better. For the enhanced EEC, with a minimum mask of 7 ns, it requires a test instrument with an accuracy of 0.7 ns or better [6].

Figure 6.13a, b illustrate MTIE/TDEV of EEC option 1 as defined in ITU-T G.8262 and MTIE/TDEV of eEEC (defined by ITU-T G.6282.1), respectively.

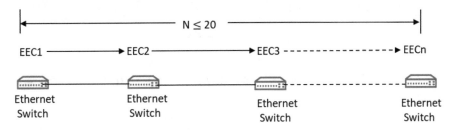

Fig. 6.11 Chain of SyncE EEC connected in a series

Native Ethernet Networks with no SyncE enable Ethernet Switch

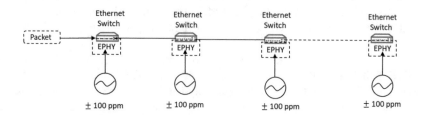

SyncE Network with SyncE enable Ethernet Switch

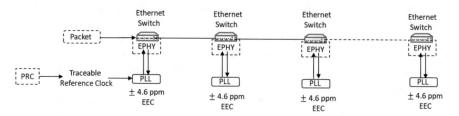

Fig. 6.12 The functional outlooks of native ethernet and SyncE-based ethernet network

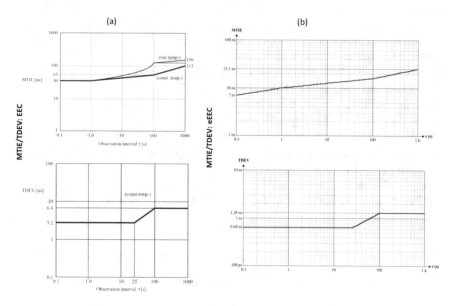

Fig. 6.13 The difference between EEC and eEEC by applying MTIE/TDEV mask (Courtesy: ITU-T G.8262 and G.8262.1)

Option 2 of EEC is not presented in the figure. The physical layer clocking as implemented in the SyncE network suffers from jitter and wander that diminishes the
quality and reliability of the clock as it passes through the network. Due to this, the
SyncE EEC has a restricted allocation of the overall system-level timing budget. In
this context, jitter is defined as phase variations above 10 Hz bandwidth and wander
is defined as the variations occurring at a frequency below 10 Hz. Using 10 Hz as
the dividing line between jitter versus wander is an arbitrary convention that has
been used in the telecom industry for many years. In truth, jitter and wander are both
phase variations, but their effects in the network are different [7].

From operational perspective, SyncE uses Digital Phase Lock Loop (DPLL)
which provides jitter and wander filtering, and also translates telecom clocks to
ethernet clocks. A PLL can be either analog or digital, the difference is that DPLL
has inherently much higher level of noise immunity than analog and hybrid analog
implementations. For the SyncE implementation, output of the DPLL is also used to
drive the clock of the Ethernet PHY (EPHY) as depicted in Fig. 6.14. This network
scenario was discussed in Chap. 2 and specified in ITU-T G.8271.1. Trimble's GM
200 is acting as T-GM providing primary reference clock traceable to UTC for
SyncE devices. All ethernet switches that downstream from T-GM is integrated with
SyncE eEEC implemented through DPLL. Node downstream can recover the PRC/
UTC traceable clock from the receive path of EPHY.

Please note communication between RU (Radio Unit) and eNodeB is also using
EPHY with DPLL due to the implementation of eCPRI which uses ethernet at physical and MAC level. At each node level DPLL helped recover clock and provides to
EPHY and this continues at network level. The SSM mechanism discussed earlier
can also be used in SyncE to convey timing quality information for the downstream
ethernet switches. If upstream network synchronization failure occurs, the synchronization function processes the SSM and takes appropriate action to select a new
timing reference. The method for processing SSM through ESMC (Ethernet
Synchronization Messaging Channel) is discussed in ITU-T G.8264. The ESMC
allows extraction of QL for reference frequency retrieval.

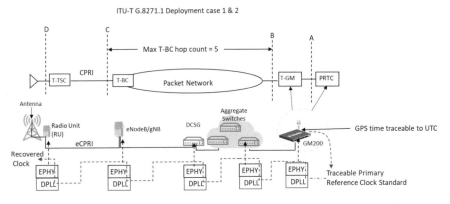

Fig. 6.14 Typical deployment of SyncE eEEC through the implementation of DPLL

References

1. Chowdhury, D. (2020). *Addressing 5G sync plane issues.*
2. G.8264. *Distribution of timing information through packet networks.* ITU-T G.8264/Y.1364 (08/2017). International Telecommunication Union.
3. RCC. (2016). *Overview of IRIG-B time code standard.* TN-102. IRIG STANDARD 200-04. Range Commanders Council (RCC).
4. IRIG. (2016). *IRIG serial time code formats.* IRIG Standard 200-16. Range Commander Council (RCC).
5. Rajput, A. (2014). Implementation of IRIG-B output for time synchronization. *International Journal of Advanced Research in Electronics and Communication Engineering (IJARECE), 3*(5) http://citeseerx.ist.psu.edu/viewdoc/download? doi=10.1.1.670.1156&rep=rep1&type=pdf.
6. Calnex. (2019). *Paragon-X: can Paragon-X be used to measure Enhanced SyncE?* Calnex Solutions plc.
7. Geber. (2020). *Maintain precise timing across large, heterogeneous SyncE networks.* AspenCore, Inc. Available at edn.com

Chapter 7
Packet Timing: Network Time Protocol

7.1 Introduction

Although PTP (Precision Time Protocol) is making increased penetration in various network infrastructure, NTP (Network Time Protocol) remains the most commonly used and widely deployed packet timing protocol till date. Globally, a big virtual cluster of public NTP servers provide reliable and easy to use NTP time services to millions of clients. The protocol is part of TCP/UDP/IP suits and uses UDP as the underlying protocol to transport NTP clock services. The NTP network is organized in stratum hierarchy levels and operates in different modes for clock exchange. Since its development, NTP went through various revisions and currently NTP version 4 is widely used. A variation of NTP deployment is Chrony daemon based NTP implementations that expected to provide higher accuracy than standard ntpd based NTP. In recent years, much improvement is done in NTP security mechanisms with introduction of Network Time Security (NTS).

This chapter discusses NTP including its protocol structure and operation with added details on newer security mechanisms for NTP such as NTS and NTPsec. Readership will find this chapter useful to understand deployment constraints and design configurations that are needed. More importantly this chapter and the proceeding chapter provide a choice for customer to choose right packet timing protocol for their network design based on appropriate time accuracy needed for respective applications.

© The Author(s), under exclusive license to Springer Nature
Switzerland AG 2021
D. D. Chowdhury, *NextGen Network Synchronization*,
https://doi.org/10.1007/978-3-030-71179-5_7

7.2 Network Time Protocol (NTP)

Till date, NTP remains one of the most commonly used packet timing protocols despite the increased penetration of PTP (Precision Time Protocol). Most computers and network devices still use NTP to retrieve and synchronize time. From historical perspective, NTP's roots can be traced back to a demonstration at NCC 79 (1979 National Computer Conference) believed to be the first public coming-out party of the Internet operating over a transatlantic satellite network. However, it was not until 1981 when synchronization technology was documented in this historic Internet Engineering Note series as IEN-173 [1]. The document attempted to present a conception towards logical clocks that can be synchronized among the DECnet hosts through routing updates as long as "oscillator drift rates are stable and do not differ radically." Created by Digital Equipment Corporation, DECnet was the most common networking protocol suit at the time and IEN-173 was an attempt to define clock synchronization among the DECnet hosts. Extending the works presented in IEN-173, the first public specification for time synchronization appeared in RFC778 [2]. Later, the first deployment of the technology in a local network was an integral function of the Hello routing protocol documented in RFC891 [3]. In 1983, J. Postel and H. Harrenstein published a two-page RFC, namely "RFC868" defining "Time Protocol"—a site-independent, machine readable date and time that could operate as a service on top of TCP and UDP protocol. The time protocol as defined in RFC868 uses port 37 for TCP and UDP to listen to requests and returns a 32-bit value of time, which is represented in seconds since 00:00, midnight, on January 1, 1900, Greenwich Mean Time (GMT). Given the capacity of a 32-bit number, this simple time protocol will likely require some adjustments by the year 2036 [4]. David L. Mills who wrote IEN-173 followed by RFC778 and RFC891, continued his works and defined NTP (Network Time Protocol) through a series of RFC for which the first was RFC 958 published in September 1985. RFC contained first formal specification for NTP version 0 with NTP packet header and offset/delay calculations, etc. Considering the speed of the network and other network transport constraints at the time, the accuracy of NTP version 0 was around 100 ms. Mr. Mils later introduced NTP version 1.0 in RFC 1059 which contained first comprehensive specification of the protocol and algorithms, including primitive versions of the clock filter, selection, and discipline algorithms [1]. David continued his works to augment NTP through a series of RFCs: RFC1119 (NTP version 2), RFC1305 (NTP Version 3), and RFC5905 (NTP version 4). Subsequently, a simplified version of NTP (SNTP) was introduced through a series of RFCs starting with RFC1361 followed by RFC1769, RFC2030, and RFC4330. These RFCs collectively defined SNTP for network scenarios where less stringent time distribution needed.

7.2.1   What Is NTP?

NTP is one of the application services protocol in the TCP/UDP/IP suit. Other application services protocols are FTP, SNMP, SMTP, and Telnet to name few. NTP uses well-known UDP port 123 and the implementation relies on UDP for transport. The fundamental purpose of NTP is to provide time synchronization among participating network devices with reference to a reliable clock source which can be either absolute or relative. The absolute clock represents GNSS/GPS time traceable to UTC, whereas relative time represents a time on a device with which all of the other networking devices must synchronize but that time is not derived from or synchronized with a UTC source. The relative time could simply be the time set by a network administrator to approximate the time of an electric clock that is located on a wall within a building where the network is deployed. Ironically, a device could be configured as a stratum 1 without deriving its time from UTC. The practice is generally not recommended, but not discouraged either as long as the network administrator is aware of its implications. The NTP is a client/server application for which NTP algorithm is implemented in computers and networking devices as part of the OS (Operating System) module.

7.2.2   Hierarchical Nature of NTP

NTP client/server application works in a clock hierarchy known as stratum. We have discussed about the stratum clock hierarchy in Chap. 3. NTP clock hierarchy is configured in a stratum level with stratum 1 as the top of the hierarchy followed by stratum 2, 3, and 4 clocks. As shown in Fig. 7.1, Stratum 1 level clock can be derived from an atomic clock or a specially designed time server with GNSS/GPS time traceable to UTC. Trimble's Thunderbolt™ TS200 is a NTP time server with GNSS time traceable to UTC acting as stratum 1 level NTP server in this network deployment scenario. Network administrator may also deploy an atomic clock with or without GNSS time reference traceable to UTC. However, TS200 type of "stratum 1" time servers are very cost-effective than atomic clock.

Core routers are configured as stratum 2 level devices providing clock input to network segments for stratum 3 level distribution as depicted in the figure above. Generally, Layer 2/Layer 3 (L2/L3) switches provides stratum 3 level time distribution for network segments. Edge devices at LAN (Local area network) interfaces of L2/L3 switch can provide stratum 4 level time input to other devices. The network scenario shown in Fig. 7.1 is a typical scenario and hence, stratum level time distribution for NTP may vary depending upon network architecture. Please note that end user devices in LAN are all NTP clients. However, it's also possible to create a collapsed flat architecture in which all networking devices in the topology are stratum 1 server.

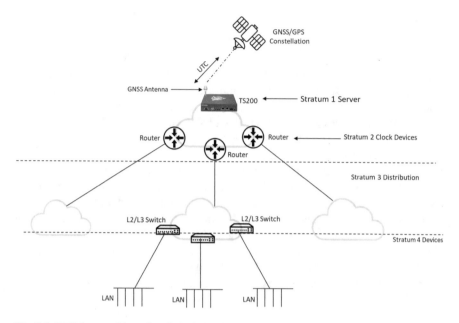

Fig. 7.1 NTP Stratum hierarchy of clock distribution

7.2.3   NTP Protocol

As discussed, NTP is part of TCP/UDP/IP applications services protocol and uses IP and UDP for connectionless transport. It was designed specifically to maintain accuracy and reliability, even when used over typical Internet paths involving multiple gateways and unreliable networks. There is no provision for peer discovery, configuration, or acquisition though some implementation includes these features. There are no circuit management, duplicate detection or retransmission facilities are provided or necessary in NTP operation. The data integrity is performed through IP and UDP checksum. NTP can operate in different modes depending upon network scenarios involving private workstations, public servers, and various network configurations. The mode of operation allows communication between two different devices that includes NTP Time requests and NTP control queries. A client uses time request to obtain time information from an NTP server. NTP control queries are the communication messages for configuration information. Figure 7.2 illustrates time request and response between stratum 1 server and stratum 2 clock.

Same method applies to a flat architecture where NTP server and clients are directly connected, eliminating the need for intermediary clock levels. Devices in a network will periodically update their clock based on NTP server response.

There are four modes of operation for NTP clock exchange:

- *Server Mode*: In this mode of operation, NTP application is configured as such that the device will synchronize NTP clients. NTP servers can be configured to

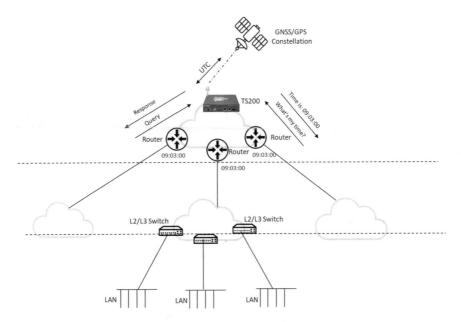

Fig. 7.2 NTP message exchange between server and client

either synchronize all clients or only specific groups of clients. NTP server will not allow clients to update or manipulate a server's time settings. It only provides time information to clients and hence, will not accept synchronization information from their clients.

- *Client Mode*: Normally, all network routers and switches are configured as NTP client to obtain clock from NTP time server and adjust device clock accordingly. It is to be noted that NTP clients can be configured to use multiple servers to set their local time and can be configured to give preference to the most accurate time sources available to them. If configured in NTP client mode, devices will not provide synchronization services to any other devices.

- *Peer Mode*: In this mode of operation, NTP-enabled device does not have any authority over another. Each device will share its time information with its peer. Additionally, each device can also provide time synchronization to the other. NTP peers can operate in symmetric active and symmetric passive modes. In symmetric active configuration, device will send synchronization request packet to the symmetric passive device with the mode field set to 1. The value "1" indicates the symmetric active mode. Once symmetric passive device received the packet, it will respond with the mode field set to 2. The value 2 indicates the symmetric passive mode. Through this process of message exchange both symmetric active and symmetric passive endpoints are synchronized where lower stratum will synchronize with higher stratum.

- *Broadcast/Multicast Mode:* This is a special server mode configuration in which the NTP server broadcasts its synchronization information to all clients.

Broadcast mode requires that clients be on the same subnet as the server, and multicast mode requires that clients and servers have multicast capabilities configured. Broadcast and multicast mode are useful in a network scenario where significant numbers of devices are configured as NTP clients since such mode of operation requires less resources.

Additionally, NTP version 4 also support manycast and anycast where manycast is an automatic discovery and configuration mechanism and anycast is directed communication between NTP servers to sync clients via a single IP address. Manycasting is intended for multicast clients to search the network neighborhood to find cooperating manycast servers. The client sends UDP packets to the configured multicast address with its TTL field set to zero, to see whether there are any manycast servers on the local segment. If the client doesn't get a response in a given period of time, it will retransmit with TTL field set to 1 and this process continues until TTL set to a maximum value and no server responds or client eventually finds a server (s) to synchronize. In manycast mode, once an NTP client finds the server(s) it establishes uncast association with them. In contrast, anycast uses a single IP address assigned to multiple servers for which routers direct packets to the closest active server. For synchronization over internet services, anycast uses known IP address such as DNS. For enterprise network configuration, network managers may decide to simplify the configuration of many NTP clients by configuring network clients with same NTP server IP address and allocating a group of anycast NTP servers to service the synchronization request. Depending upon network configuration and deployment scenarios, network manager may choose to use certain NTP mode of operation.

Table 7.1 summarizes network scenarios in terms of mode categories and presents corresponding mode of operation for each. Based on the table provided network managers can decide appropriate configuration for NTP operation in their networks.

Table 7.1 NTP Mode categories and corresponding mode of operations

Mode categories	Mode of operation
Point to point	Peer mode: Symmetric active
	Peer mode: Symmetric passive
	Client
	Server
Point to multipoint	Broadcast
	Multicast
Multipoint to point	Manycast/Anycast

7.2.3.1   NTP Message Structure

The message sent by NTP clients to the server includes an UDP encapsulated NTP packet with 17 different fields as depicted in the figure below. The packet has 12 words followed by optional extension fields and finally an optional message authentication code (MAC) consisting of the Key Identifier field and Message Digest field [5]. The extension fields may be used to add optional capabilities, e.g., Autokey security protocol defined in RFC5906. The MAC is used for both Autokey and the symmetric key authentication. The packet starts with LI (Leap Indicator), a 2 bits integer field warning of an impending leap second that can be inserted or deleted to UTC timescale at the end of the current month. The values for this field are: 0 for warning, 1 for last minute of the day has 61 s, 2 for last minute of the day has 59 s, and 3 for unknown or clock is unsynchronized. This field is followed by VN (Version Number), a 3-bit integer field that indicates NTP version number, e.g., 4 for NTP version 4. The next field is "Mode," a 3-bit integer describing NTP mode of operation which is discussed earlier. NTP version 4 RFC5905 defines following values for mode field: value of 0 is reserved, 1 for symmetric active, 2 for symmetric passive, 3 for client, 4 for server, 5 for broadcast, 6 for NTP control message, and 7 is reserved for private use. The next field after "mode" is "stratum," an 8-bit integer representing the stratum as depicted in Fig. 7.3.

The values for stratum field are: 0 for unspecified or invalid condition, 1 for primary NTP server equipped with GPS receiver, e.g., Trimble's TS200 as depicted in Fig. 7.2. The value of 2–15 in the stratum field represents secondary server (via NTP) while value of 16 indicates unsynchronized state and 17–255 is reserved for future use.

Next is an 8-bit signed integer field, "poll" that indicates interval between successive messages in seconds sent by the NTP peers. Each server uses the minimum

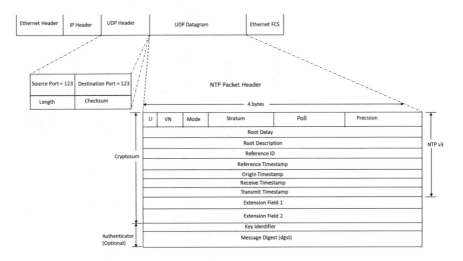

Fig. 7.3 NTP version 4 message structure [5]

of its own poll interval and that of the peer. Suggested values for the field are 6 for minimum interval and 10 for maximum interval. To indicate the precision of the system clock, an 8-bit signed integer field "precision" is used. For example, a value of -18 indicates a precision of about one microsecond. Round trip delay and total depression to the reference clock are indicated in root delay and root depression field. The reference ID field, a 32-bit code identifies the particular server or reference clock. For stratum 1, the field contains 4 bytes left justified ASCII string. Table 7.2 depicts ASCII identifiers for the field as per RFC5905.

Reference timestamp field indicates the last update and is used for management purpose, whereas original timestamp is the local time at which the client sent a request message to the server, in the same format as the reference timestamp. The transmit and destination timestamp indicate time at server during the response and time when message arrived at the client, respectively.

The MAC (message authentication code; not to be confused with ethernet MAC) consists of "key identifier" followed by message digest. The message digest or checksum is calculated overall NTP header including extension fields. The MAC field is also known as authenticator. It is optional and only present when using authentication. The key ID is used by the client and server to designate a secret 128-bit MD5 key.

Table 7.2 ASCII string for reference ID field [5]

ID	Clock source
GOES	Geosynchronous orbit environment satellite
GPS	Global position system
GAL	Galileo positioning system
PPS	Generic pulse-per-second
IRIG	Inter-range instrumentation group
WWVB	LF radio WWVB Ft. Collins, CO 60 kHz
DCF	LF radio DCF77 Mainflingen, DE 77.5 kHz
HBG	LF radio HBG Prangins, HB 75 kHz
MSF	LF radio MSF Anthorn, UK 60 kHz
JJY	LF radio JJY Fukushima, JP 40 kHz, Saga, JP 60 kHz
LORC	MF radio LORAN C station, 100 kHz
TDF	MF radio Allouis, FR 162 kHz
CHU	HF radio CHU Ottawa, Ontario
WWV	HF radio WWV Ft. Collins, CO
WWVH	HF radio WWVH Kauai, HI
NIST	NIST telephone modem
ACTS	NIST telephone modem
USNO	USNO telephone modem
PTB	European telephone modem

In NTP version 4, one or more extension field can be inserted after the header and before the MAC. Other than defining this optional field, RFC5905 does not state the use of extension field. However, RFC7822 provides updates to RFC5905 regarding extension fields. According to RFC 7822, an extension field may be used by the MAC, e.g., autokey as defined in RFC5906. Autokey is defined as security model for authenticating servers to clients using the NTP and public key cryptography. Please refer to RFC5906 for further details to learn more on autokey. If the extension field requires MAC as per the implementation of a specification like autokey, the extension field must specify the algorithm used. An extension field may allow the use of more than one algorithm, in such case it should include information about those algorithms.

7.2.4   Approach to NTP Deployment

There are four steps to designing an effective NTP deployment and those are: choosing the time source, NTP topology, determining NTP features to configure and monitoring and management of NTP operations. Each step outline above offers several choices and user should weigh pros and cons of each options before deploying. The following four steps identify some guidelines for successful deployment of NTP in a given network.

Step 1: Primary Clock Source
Given the available choices for NTP time source, this step sets the tone for the rest of the design process, as it will impact the topology, configuration, and management aspects of NTP deployment. The NTP time source choices include:

- One or more private NTP stratum 1 timeservers with a primary reference clock traceable to UTC via GPS, CDMA, radio, or modem.
- Public stratum 1 NTP servers available over the Internet. If an enterprise is building capabilities and services that are intended to be deployed outside of the enterprise, network administrators should consider requesting appropriate public NTP server for the region from the pool of available servers at https://www.pool.ntp.org/ . It is important to note that most public NTP servers specify route of engagement. Enterprise should set up hierarchy before allowing devices to access public NTP server.

Step 2: Topology and NTP Hierarchy
It is possible to purchase NTP stratum 1 appliance to use internally for less than the cost of a typical server. A user may also choose to build their own NTP stratum 1 server; however, one must consider the following best practices irrespective of whether it is purchased on built-in house:

- *Standardize to UTC time:* The primary NTP server must provide primary reference clock traceable to UTC time. Also, network administrator should consider standardization of all systems to UTC within an enterprise, as standardizing to

UTC simplifies log correlation within the organization and with external parties no matter what time zone the device being synchronized is located in.

- *Number of NTP clients*: Determining number of NTP clients may help network administrator to choose right stratum 1 level NTP server and consider the need for the presence of secondary, higher-stratum servers for the network.
- *Segal's Law*: Remember segal's law, "a man with one watch knows what time it is. A man with two watches is never sure." Well, it is actually work to the contrary. In fact, a man with one watch can't really be sure he knows the right time, he merely has no way to identify error or uncertainty. The said is true for the NTP deployment. It would work best to have three or more stratum 1 servers and use those servers as primary masters. Three or more stratum 1 server would provide a more accurate timestamp because they are using a time source that is considered definitive.
- *Redundancy*: Network infrastructure redundancy must be considered for NTP deployment as this will impact the NTP topology itself.

Step 3: NTP Features

Network administrator should consider mode of NTP operation preferred for their network deployment aforehand to avoid misconfiguration of NTP parameters. For example, if network prefers to avoid unnecessary traffic and direct NTP clients to pull of server, manycast could be appropriate than other form of NTP server discovery by the NTP clients. Other consideration includes the following:

- *Basic features*: A more typical configuration would include elements of some or all of the optional features and capabilities that are listed next.
- *Security features*: Enterprise should consider two distinct areas of focus while implementing security measure for NTP deployment: first securing NTP stratum servers. It is recommended to restrict the commands that can be used on the stratum servers. Network administrator must not allow public queries of the stratum servers. Only known networks/hosts should be able to communicate with their respective stratum servers. Secondly, network administrator should consider the business need for cryptography. Given increased level of hacking in enterprise networks, it is recommended to use encrypted communications and encrypted authentication. However, one note of caution is that cryptographic services associated with NTP for securing NTP communications require key management and put higher computational overhead of servers.
- *Operational Modes*: As discussed earlier, operational mode is an important consideration as it impacts overall NTP deployments. Careful consideration is needed to avoid network overhead since each NTP mode of operation has its own pros and cons. Thus, understanding network underlay is also important.

Step 4: Monitor and Manage NTP Operations

The approach to NTP management will vary depending upon objectives, for example some network administrator may prefer to monitor devices and servers connected including the alarms, logs, etc., while others may consider detection of attack as part of monitoring and management of NTP operation. It is recommended that network administrator should monitor NTP instances to detect attacks in addition to standard alarms and logs. Many known attacks have particular signature, e.g., bogus packet whose timestamp origin does not match the value expected by the client. An IETF draft at https://tools.ietf.org/id/draft-ietf-ntp-bcp-08.html identifies some of the best practices for operation and management of NTP deployments. It is recommended to consider guideline stated in the IETF draft "Network Time Protocol Best Current Practices."

7.3   Simple Network Time Protocol (SNTP)

A series of RFCs starting with RFC1361 in 1992, RFC 1769 in 1995, RFC 2030 in 1996, and RFC 4330 in 2006 defined SNTP. It is to be noted that NTP version 4 RFC5905 obsoleted SNTP version 4 RFC 4330. The release of SNTP in 1992 was primarily due to the limited computing power of the computers available at that time. The SNTP used less processing power during synchronization than NTP, which frees up the processor for other tasks. However, most modern devices have a processing power which can accommodate the complexity of NTP. Both NTP and SNTP are very similar applications services protocol of TCP/IP suits in that they use the same time packet from a time server message to compute accurate time. The methods used by the time servers to assemble and send out a timestamp are exactly same whether NTP or SNTP is used for synchronization. A NTP client may choose to use NTP or SNTP depending upon program running on them but primary time servers do not care type of program running at the client. The difference between NTP and SNTP can be distinguished according to the following points:

- Number of servers used for the synchronization process.
- Number of algorithms used to make up for time deviations and ensure the most accurate results possible.

In contrast to NTP, SNTP prioritizes simplicity. It suggests that the time information should only be obtained from a single server, and any further client-server dependency should be avoided. NTP on the other hand relies on a complex construct of different servers that pass on the information in a hierarchical layer system. In comparison to NTP, SNTP contains fewer algorithms. SNTP does not use some algorithms that are provided as standard in the NTP specification. NTP algorithms are much complex than SNTP since NTP normally uses multiple time servers to verify the time and then controls the slew rate of the NTP clients. The algorithm determines if the values are accurate using several methods, including fudge factors and identifying time servers that don't agree with the other time servers. NTP algorithm then speeds up or slows down the client clock's drift rate so that the NTP client's time is always correct and there won't be any subsequent time jumps after the initial correction. SNTP does not support such algorithm as a result, it offers a lower level of accuracy than NTP, making the SNTP unsuitable for applications and processes that require incredibly accurate time synchronization. However, due to its use of fewer algorithms and simplified client-server communication, SNTP requires much fewer resources than NTP, which is particularly helpful for simple devices or systems with low computing power [6, 7].

7.4 Chrony

Recently Facebook published an article depicting reasons as to why they have moved to Chrony daemon based NTP implementation from standard NTP daemon (ntpd). The reasoning provided in an article states, "Facebook needed to sync all the servers across many data centers with sub-millisecond precision and for this reason we found that chrony is significantly more accurate and scalable than the previously used service, ntpd, which made it an easy decision for us to replace ntpd in our infrastructure." [8].

Chrony is an open-source initiative that replaces ntpd with chrony daemon for NTP implementation and adheres to NTP version 4 (RFC5905). It can synchronize the system clock with NTP servers, reference clocks (e.g., GPS receiver), and manual input using wristwatch and keyboard. Chrony is designed to perform well in a wide range of conditions, including intermittent network connections, heavily congested networks, changing temperatures and systems that do not run continuously, or run on a virtual machine [9]. According to Facebook experiment [8], Chrony helped the organization to achieve an accuracy of 100 microsecond from 10 ms for the NTP implementation. It is anticipated that a sub-microsecond accuracy is achievable with hardware timestamping. Additionally, chrony has features that make it the better choice for most environments for the following reasons:

- It can synchronize to the time server much faster than NTP.
- Chrony can compensate for fluctuating clock frequencies, e.g., when a host enters sleep mode, or when the clock speed varies due to frequency stepping that slows clock speeds when loads are low.
- It can handle various network conditions, such as intermittent network connections and bandwidth saturation, delays, and latency.

Chrony is distributed under GNU public General Public License version 2. The distribution is available in many linux operating systems such as Fedora, Ubuntu, RHEL, CentOS, etc.

There are two programs in chrony:

- chronyd—a daemon that can be started at boot time,
- chronyc—a command-line interface program that can be used to monitor chronyd's performance and to change various operating parameters while it is running.

NTP client configuration requires little or no intervention. The default configuration file is /etc./chrony.conf file. For further details on chrony and server configuration, please refer to respective linux distribution userguide.

7.5   NTS and NTPSec

The NTS (Network Time Security) is the latest attempt to secure NTP using crypto-graphic protection while NTPSec (Secure Network Time Protocol) is an open-source project that uses NTS for NTP implement in linux systems. RFC8915 that released in September 2020 defines security mechanism for client-server mode pro-posing NTS as an extension field for the NTP version 4. It uses Transport Layer Security (TLS) version 1.3, Authenticated Encryption with Associated Data (AEAD), and digital certificates. Evolved from SSL (Secure Socket Layer), TLS is considered improved version of SSL for encryption, authentication, and integrity. It works same way as SSL. Defined by RFC8446, TLS 1.3 is considered seventh itera-tion of SSL/TLS and offers myriad improvements over its predecessors, including a new handshake and revamped cipher suites. It uses AEAD algorithm instead of symmetric encryption algorithms that are available in TLS 1.2. The AEAD is a form of encryption that simultaneously assures the confidentiality and authenticity of data.

Apart from cryptographic protection, objectives of NTS are to provide confiden-tiality, replay-prevention, request-response consistency, unlinkability, non-amplification, scalability, and performance. For further details on these features, please refer to Sect. 1.1 of RFC8915. The NTS uses TLS 1.3 to establish keys, to provide the client with an initial supply of cookies, and to negotiate some additional protocol options as part of "NTS Key Establishment" (NTS-KE) mechanism. In this process, client connects to an NTS-KE server on the NTS TCP port 4460 and two parties perform a TLS handshake as shown in Fig. 7.4. Via this TLS connection, client and server negotiate some additional protocol parameters, and the server sends the client a supply of cookies along with an address and port of an NTP server for which the cookies are valid.

Client's initial TCP connection to NTS-KE server is released once parameter exchange is complete and key establishment procedure is done. The communica-tions between NTS-KE server and NTP server to exchange shared encryption

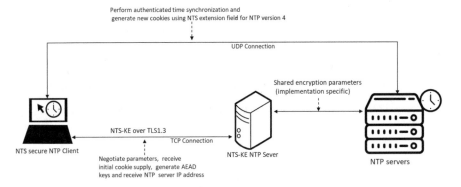

Fig. 7.4 NTS connectivity for NTP version 4

parameters are not defined in RFC8915 and left to the vendor specific implementations. Client establishes a connection with NTP server on received IP address and UDP port after NTS-KE server negotiation is done. The default UDP port for UDP communication with NTP server is port 123. The negotiation between NTPv4 server and client is described in RFC8915 section 4.1.7. Readership may refer to this section of the RFC8915 for further details.

As discussed, the NTS mechanism is also used by NTPsec for implementation of secure NTP in linux environment and managed by NTP project at https://www.ntpsec.org/ . It is an open-source implementation for Linux that tries to secure NTP and not related to the release of NTS. At this writing, the latest version of NTPSec is 1.1.9 and available for many linux distribution such as Ubuntu, Alpine, Debian, Fedora, FreeBSD, and Gentoo.

Readership should consult respective linux distribution websites for details instruction about the installation of NTPsec. For Ubuntu, once can start with running update command followed by installation of NTPsec program as follows:

sudo apt-get update -y
sudo apt-get install -y ntpsec

For Ubuntu 20.10, you may need to add the Universe repository to the sources of APT: *sudo add-apt-repository universe*. After adding the repository, you can install NTPSec: *sudo apt install ntpsec*.

References

1. Mills, D. L. (1981). *IEN-173: time synchronization in DCNET hosts*. COMSAT Laboratories. Available online at https://www.eecis.udel.edu/~mills/database/rfc/ien-173.txt
2. Mills, D. L. (1981). *RFC778: DCNET internet clock service*. COMSAT Laboratories. Available online at https://www.rfc-editor.org/rfc/rfc778.txt
3. Mills, D. L. (2002). A brief history of NTP time: confessions of an internet timekeeper. *ACM SIGCOMM Computer Communication Review, 33*, 9–21.
4. Rybaczyk, P. (2005). *Expert network time protocol: an experience in time with NTP*. APRESS.
5. RFC5905. (2010). *Network Time Protocol Version 4: protocol and algorithm specification*. Internet Engineering Task Force (IETF).
6. SNTP. (2018). *Simple network time protocol: the stripped-back protocol for time synchronization*. Digital Guide IONOS by 1&1.
7. Orolia. (2019). *Technical note: differences between NTP and SNTP*. Orolia.com
8. Obleukhov, O. (2020). *Building a more accurate time service at Facebook scale*. FACEBOOK Engineering. Available online at https://engineering.fb.com/2020/03/18/production-engineering/ntp-service/ `
9. Chrony. (2019). *Chrony introduction*. TUXFamily.org. Available online at https://chrony.tuxfamily.org/

Chapter 8
Packet Timing: Precision Time Protocol

8.1 Introduction

The Precision Time Protocol (PTP) is one of the important packet timing protocol for next generation network synchronization. It is a packet-based two-way communications protocol specifically designed to precisely synchronize distributed clocks to sub-microsecond resolution, typically on an Ethernet or IP-based network. Defined by IEEE1588 standards, PTP provides real-time applications with precise time-of-day (ToD) information and timestamped inputs, as well as scheduled and/or synchronized outputs for a variety of systems in different industry specific networks, ranging from LTE/5G based mobile networks, industrial automation, audio-visual networks, smart grid to transportation, automotive and Industrial IoT networking. The PTP operation is based on master-slave hierarchy in which PTP network is self-organized to synchronize all clocks to grandmaster clocks in single or multiple domains. PTP defines a number of clock categories and uses best master clock algorithm to select master clocks for the network. Given the need for real-time and near real-time traffic more most modern network design, PTP offers high-precision accuracy for time distribution ideal to time-sensitive real-time and near real-time traffic. Additionally, PTP version 2.1 provides internal and external security mechanisms for secured time distribution across various industry specific network. In this chapter a detail of PTP protocol architecture, profiles, operation, transport, and security mechanism are discussed. Readership will find this information useful in the design and optimization of next generation networks that support time-sensitive traffic.

© The Author(s), under exclusive license to Springer Nature
Switzerland AG 2021
D. D. Chowdhury, *NextGen Network Synchronization*,
https://doi.org/10.1007/978-3-030-71179-5_8

8.2   Precision Time Protocol (PTP)

The 4th Industrial Revolution is rapidly changing the way we live, work, and relate to one another. The adoption of always connected services, cyber-physical systems, and IOT is blending the physical and digital worlds. Fueled by ubiquitous connectivity and computing, these interconnected systems are retrieving and processing petabytes of data. They allow businesses to gain real-time insight and provide for high degrees of personal data consumption. They are creating an era of "Intelligence." This fusion of the digital and physical worlds is pushing the development and deployment of the ubiquitous connectivity infrastructure transforming all industry verticals in process. These transformations require processing of real-time and near real-time processing of data at the edge of the network instead of transporting data back to a data center, making the computation, and then sending it back to the edge computer. Furthermore, the demand for intelligence and better business insights is pushing computing and applications to process and present data immediately. It is driving many use cases across industries—a few examples are Edge cloud, Autonomous Vehicles, Vehicle-to-Infrastructure (V2X), ATM security, credit card fraud detection, automated manufacturing, and more. All these applications are transactional in nature and need tight time synchronization. Often than not, tighter synchronization at microseconds level is required in many applications across different industries, e.g., SmartGrid, Telecommunications networks, machine vision and industrial automation, etc. In some cases, application requires nanosecond-level accuracy. These increased demand for tighter time synchronization makes NTP less effective in some of the modern communications infrastructure.

Henceforth, Precision Time Protocol (PTP) was developed to fill the void. Defined by IEEE 1588 standard, PTP distributes precise time with better than 1-μs accuracy over Ethernet. Table 8.1 depicts the difference between NTP and PTP.

It is to be noted that to achieve the sub-microsecond level accuracy and precision specified in the PTP standard, a thorough understanding of the protocol and adherence to specific design principles are needed. It is equally important to also

Table 8.1 A comparative outlook between NTP and PTP

Time distribution methods	NTP	PTP
Physical layer	Ethernet	Ethernet
Model of operation	Client-server	Master-slave
Synchronization accuracy	1–100 ms	100 ns to 1 μs (sub-microseconds)
Compensation for latency	Yes	Yes
Update interval	Minutes	Configurable typically once per second
Hardware requirements	Master only	PTP support hardware timestamping for higher accuracy
Relative cost	Low	Low to medium

understand how network asymmetry, message delay variations, and network topology affect accuracy and precision [1].

Defined by IEEE 1588 standard, PTP is a message-based time transfer protocol that provides a way to easily synchronize clocks in a distributed system, thereby enabling end-to-end time synchronization and system-wide accurate and precision timescale. PTP was first defined in IEEE 1588-2002 standard and considered version 1 of the protocol. Subsequently, PTP version 2 defined in IEEE1588-2008 was released in 2008. Current PTP Version is 2.1 is defined by IEEE1588-2019. The IEEE 1588 original works started in the 1990s at the central research laboratories of the Hewlett-Packard Company, and continued at Agilent Technologies after the split from Hewlett-Packard in 1999. It was intended for use in instrumentation systems using network communication for control and data transport. During early presentation, the technology garnered significant interest from the industrial automation community. By fall 2000, it was evident that the technology needed standardization and thus picked up by IEEE and standard works begin in the spring of 2001. The work at IEEE concluded with publication of first release of the standard in November 2002. The standard was also approved by IEC (International Electrotechnical Commission) and incorporated in IEC 61588 standard.

Please note PTP version 1 and 2 are not compatible and should not be mixed in a network environment. However, PTP version 2.1 is backward compatible to version 2. Additionally, PTP version 2.1 also introduces various security features for the protocol.

8.2.1 Fundamental Operation of the Protocol

PTP is designed for time synchronization of distributed system in a packet-based network such as Ethernet. It is distributed protocol in that each IEEE 1588-enabled device in the network executes exactly the same protocol. There is no need to configure nodes prior to operation, assuming that the default values of PTP parameters are instantiated in all IEEE 1588-enabled devices. The entire operation of protocol is implemented using information obtained through PTP message exchanges between devices. There are five operational features of PTP protocol that together allow the synchronization of clocks in a system. These features also provide sufficient management capability to observe and tailor the system to meet specific application needs. These features are:

- Setting up boundaries and communications for the system to be synchronized.
- Creating a master-slave synchronization hierarchy.
- Establishing orderly startup and reconfiguration of the system.
- Delivering the necessary information to allow slave clocks to correctly synchronize to their master.
- Offering system and clock management capability when needed by an application.

Additionally, PTP version 2.1 defines five different types of devices for communication in PTP-based packet network synchronization. These devices are ordinary clock, boundary clock, end-to-end transparent clock, peer-to-peer transparent clock, and management nodes. It is to be noted that users must consider definition of clocks for packet network presented in Chap. 5 as defined in ITU-T recommendations G.8266/Y.1376 and G.8265.1 for grandmaster clock or T-GM, G.8273.2/Y.1368.2 for boundary clock (T-BC) and slave clock (T-TSC), and G.8272.1/Y.1367.1 for ePRTC. As we apply these different kinds of clocks to various NextGen networks synchronization in proceeding chapters, concept presented in Chap. 5 for ePRTC, T-GM, T-BC, and T-TSC along with clock definitions of IEE1588-2008 and IEEE1588-2019 standard will be important. Understanding how these standards relate in terms of clock definitions will be of utter importance in determining packet-based timing deployment for future of networks. As evident from PTP version 2.1 device types, IEEE1588-2019 exclusively does not define grandmaster clock and slave clock as the device types yet it referred both grandmaster and slave PTP instances throughout the standard. Henceforth, proper interpretation of IEEE1588 standard is imperative. The ordinary clock defined in IEEE1588-2008 and IEEE1588-2019 has two states of the clock, one is "grandmaster" state, and the other is "slave" state. An ordinary clock according to IEEE1588 standard is a device that either serves time or synchronizes to time and communicates on the network through a single PTP port, i.e., Ethernet interface. It is called a grandmaster clock if it is providing the time to entire PTP network and thus is the primary source of time for all other devices in the network or section of the network. In such case, ITU-T Recommendation G. 8266 applied to the ordinary clock device behaving as a master clock. In a real-world situation, user will find use of two different kind of master clocks providing primary reference clock source for the network, one is T-GM and the other is ePRTC. While T-GM is generally used as primary reference clock for section of network, ePRTC acts as primary reference clock for the core of the network. The figure below depicts how an ePRTC and T-GM can be placed in a network. For this example, a Telecom network is considered for which ePRTC is placed in the core of the network and T-GM is placed in the edge of the network. Trimble's GM200 is acting as T-GM for edge cloud serving fronthaul endpoints. In this case, DCSG (Disaggregated Cell Site Router) or CSR (cell Site Router) is acting as T-BC. The DCSG devices have built-in T-BC subsystem allowing them to function as boundary clock and getting the primary reference clock from T-GM, herein GM200 as shown in Fig. 8.1.

Both Radio Unit (RU) and eNodeB may implement slave clocks (T-TSC) as part of their system and thus synchronizing with DCSG. For these slaves, DSCG is acting T-BC providing master clock through its ethernet ports and also synchronizing with T-GM for primary reference clock. T-GM also synchronizes with other T-GMs and ePRTCs to establish primary reference source respective to their clocks.

As discussed, boundary clocks or T-BCs like DCSGs are multiport network device that synchronizes to the reference time on one port and serves time on one or more ports. In this case, one of the ports is a slave port (the port that synchronizes with T-GM) and the rest of the ports are master ports. In essence, a T-BC terminates

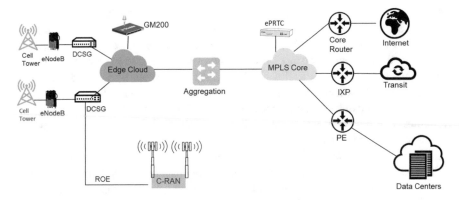

Fig. 8.1 Placement of T-GM and ePRTC for Telecom network

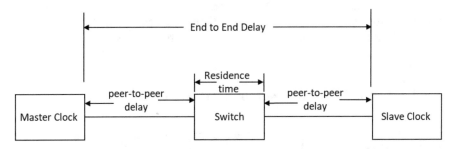

Fig. 8.2 Example of end-to-end delay and peer-to-peer delay as applicable to transparent clock

and then start the time distribution. Boundary clocks are typically used (in most cases implemented through switches and routers, however, standalone boundary clocks are also available) to scale up PTP network by servicing requests from slave clocks which would otherwise be serviced by T-GM.

An end-to-end transparent clock is a multiport network device that measures the length of time a PTP message spends within the device as it is routed from the ingress port to the egress port and then adds that information to a correction field in the message. This is intended to eliminate any variations in message delays and asymmetry that the device may introduce in the transfer of PTP messages. Unlike T-BC and ordinary clock, end-to-end transparent clock does not maintain port state, e.g., master or slave.

Similarly, a peer-to-peer transparent clock measures the link delay of each port and adds that information and the residence time to PTP messages traversing the device. It also eliminates asymmetry and packet delay variations in the device. Additionally, peer-to-peer transparent clock allows for scaling because slave devices do not have to send requests to the grandmaster clock to measure the end-to-end delay. Figure 8.2 depicts the relationship between the end-to-end and peer-to-peer delay measurements. For a network with redundant paths, peer-to-peer delay measurement is best since the PTP message will always contain the actual delay it experiences on the network, regardless of the path it takes. On the other hand, end-to-end delay measurement is useful where propagation path has PTP-unaware switches or routers. It is to be noted that implementation of peer-to-peer delay or end-to-end delay mechanism solely depends on network scenarios. Peer-to-peer delay mechanism will be useful if all switches in between master and salves are 1588 capable. If not, end-to-end delay mechanism should be used.

The next PTP device type is management node that simply used to configure and monitor PTP device. It may be a network connected computer with PTP management tools. More often, PTP grandmaster and standalone boundary clocks can be managed through either web-based user interface or various APIs such as REST and NETCONF APIs. In this case management node can be any computer with a browser or appropriate management tool for retrieving information using REST and NETCONF APIs. Linux PTP software, an open-source PTP client however support open-source PTP management client (PMC). This software can be installed in a management node monitor PTP client. Information such as number of communication paths to grandmaster, clock offset, estimated delay of synchronization and whether PTP clock is synchronized to master clock can be obtained using PMC.

8.2.2 Network Implementation Assumptions

IEEE1588-2019 standard specifies certain assumption and recommendations for the PTP in a given network. Here is the list of assumptions and recommendations as stated in IEEE1588-2019 standard:

- *Cyclic Forwarding*: Network should eliminate cyclic forwarding especially in the case of spanning tree protocol (Spanning tree is a protocol in which switches exchange BPDUs messages to eliminate loop when multiple paths exist). Though PTP eliminates cyclic forward in PTP messages, one must not take it for granted and hence ensure consider eliminating cyclic forwarding in their network.
- *Network Anomalies and Packet drop*: PTP is tolerant to missed, duplication, or out of order PTP messages, however as such is considered a rare occasion. Henceforth, it is recommended that network designer consider minimizing packet drop or similar condition in the network.
- *Unicast/Multicast situation*: PTP supports both unicast and multicast and a mix (hybrid) configuration thereof. In case of multicast, network must ensure that appropriate multicast arrangement is done for PTP messages to reach respective endpoints. The same is true for unicast or hybrid scenarios.
- *Network setup for two-step PTP messages*: PTP supports one-step and two-step methods in which slave receives either one or two messages respectively from the master before it can send its own time. In cases where two-step PTP configuration is used, network must be designed as such that general PTP messages (Sync and Follow_Up) take same path. Failure to do so may cause unnecessary jitter and wander to the environment which may be undetectable.

- *Number of PTP Instances*: PTP assumes that the number of boundary clocks forming the master-slave synchronization hierarchy from the Grandmaster PTP Instance to any Slave PTP Instance is less than 255.
- *Priority treatment for PTP messages*: PTP-unaware devices may introduce jitter and wander degrade time transfer accuracy. Hence, it is recommended that network designer assign priority treatment for PTP event message. In addition, transparent clock and boundary clock should be used to minimize such condition.
- *PTP Domain*: PTP supports multiple domains for a given network. All PTP messages, data sets, state machines, and all other PTP entities are always associated with a particular domain. IEEE1588-2019 states that time established within one domain is independent of other domains.

8.2.3   PTP Operation

One of the primary objectives of PTP is to achieve microsecond-level accuracy or better. To achieve this, the protocol must provide:

1. *Event messages*: One or more events that can be timestamped and used for clock correction. PTP defines a number of protocol messages and classifies each as an event message or a general message. An event message is a message that must be accurately timestamped at the time of transmission, reception, or both [2].
2. *Communication of Timestamp*: Communicate the timestamps to the clock requiring this information.
3. *Ability to overcome timing impairment*: Various systems within a network and components within a system may introduce timing impairment and hence, PTP should overcome such timing impairments.

The PTP event message stated above consists of Sync, Delay_req, Pdelay_Req, and Pdelay_Res. The general message includes Announce, Follow_Up, Delay_ Resp, Pdelay_Resp_Follow_Up, Management, and Signaling. Management messages are used by management nodes to configure and retrieve configuration information from PTP devices. Signaling messages are used by PTP clocks to negotiate certain optional services, such as unicast transmission. Please note that by default, PTP uses multicast for communications.

The operation of PTP can be divided in two phases:

- *Phase 1*: In this phase, PTP is self-organized into a hierarchy where grandmaster clock gets the highest priority and slave clock is at the lowest level. Boundary clock and transparent clock get middle of the hierarchy level.
- *Phase 2*: PTP protocol messages are exchanged in phase 2 to synchronize all clocks with the grandmaster clock.

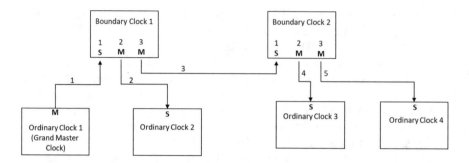

Fig. 8.3 Master-slave hierarchy selection as stated in IEEE1588-2019, clause 6

8.2.3.1 Establishing Master-Slave Hierarchy

Each port of ordinary and boundary clock has a PTP state machine that uses Best Master Clock Algorithm (BMCA) to establish master between the path for two ports and allows slave clocks use the best (most precise) time available on the network. IEEE1588-2019 standard describes operational perspectives of BMCA. Ports of ordinary clock (except slave only) and boundary clock transmit "Announce" message with clock priority and quality. Using BMCA, each clock in the network retrieves and analyzes contents and dataset from "Announce" message to determine port state which is usually master, slave, or passive.

As shown in Fig. 8.3, ordinary clock 1 is at the root of the hierarchy and thus can be denoted as grandmaster clock. Port 1 of boundary clock 1 is in slave state indicating that its local clock synchronizes with grandmaster clock. All other ports of boundary clock 1 are in master state indicating that all other devices synchronize with the local clock of boundary clock. Thus, boundary clock 2 is the slave instance to boundary clock 1 and so forth.

Only ordinary clock and boundary clock maintain this form of state and only boundary clock establishes branch point to grandmaster clock. It is to be noted that Path 1, 2, 3, 4, and 5 may have transparent clock but PTP instances in those devices do not participate in the master-slave hierarchy. ITU-T G.8265.1 (known as telecom profile) also defines a variation of BMCA operation denoting it as alternate BMCA. It is possible to have multiple masters in a given network for which each is alone in its own PTP domain as depicted in Fig. 8.4. This PTP domain separation is ensured by the network. To ensure proper separation, each master should be configured with PTP domain number. In this scenario, there will be one master active per PTP domain but all masters should be active in the network. The alternate BMCA (ABMCA) is static configuration as per G.8265.1, in which telecom slave will choose masters from a list of masters in the network. As discussed, each master is isolated by a separated PTP domain that is done through the unicast communication.

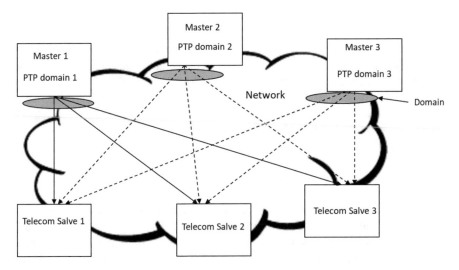

Fig. 8.4 Multiple PTP masters in a network for which each is active in its own PTP domain

The ABMCA has the following characteristics:

- Grandmasters do not exchange Announce messages.
- Masters are always active.
- Slaves are always slave-only clocks.

Since Telecom network uses redundancy, ITU-T G.8265.1 defines "telecom slave" (as depicted in the figure above) which consists of one or multiple PTP slave-only ordinary clock instances; this allows a slave node to listen to several grandmasters and any of them can be selected as grandmaster. Unlike the IEEE 1588-2008 BMCA, the ABMCA defines a fixed role for each PTP entity, e.g., T-GM, T-BC, and T-TSC. The IEEE 1588-2008 ordinary clocks which may become masters or slaves depending on the result of the BMCA are not allowed within the ITU-T G.8275.1 framework and if G.8265.1 telecom profile is selected for implementation. The purpose of the ABMCA is to let the slave clocks decide which grandmaster to use and to allow for a dynamic, loop-free architecture. With this objective in mind, the ITU-T phase/time protocol defines a new port-specific attribute, not slave, that is set to true in the T-GM, false in the T-TSC and configurable to true or false in the T-BC.

Additionally, the ways priorities are managed by the ABMCA also differ from the IEEE 1588-2008/2019 BMCA. Priority 1 is not used; it is statically configured to 128, whereas priority 1 is used in IEEE1588-2019 by the ordinary clock and its value can be 0–255. There is a new port-specific attribute, local priority that is associated with the entry of grandmasters in the telecom slave's PTP SOOC (slave-only ordinary clock) instances. The local priorities are used for master clock selection process, but its provisioning is done through management control. Local priority is locally significant and not appended in "Announce" message. Actually, the master selection process in ABMCA is based on locally provisioned priority of grandmasters in conjunction with quality level and packet timing fail attributes (PTSF-lossSync, PTSF-lossAnnounce, PTSF-unusable). In this case clockClass attributes which carry quality information have more weight and appended in the "Announce" message. The algorithm selects highest quality level that is not affected by signal fail conditions for which PTSF-lossSync is lack of reception of timing message from master, PTSF-lossAnnounce is lack of reception of Announce message from master, and PTSF-unusable is unusable timing message received by the slave.

8.2.3.2   PTP Synchronization

The following diagram shows typical message exchanges between master and slave during PTP synchronization. The main purpose of PTP is to synchronize time between two clocks and foundation of PTP for this purpose is path delay calculation. During synchronization, master clock sends timestamp to slave clock using one of the two methods: (1) One-step and (2) two-step.

The path delay calculation for these methods of synchronization is done using delay request-response (please refer to Fig. 8.5a) or the peer delay mechanism (please refer to Fig. 8.5b).

The delay request-response mechanism measures the end-to-end delay, while the peer delay mechanism measures the peer-to-peer delay. The time references in the diagram above are:

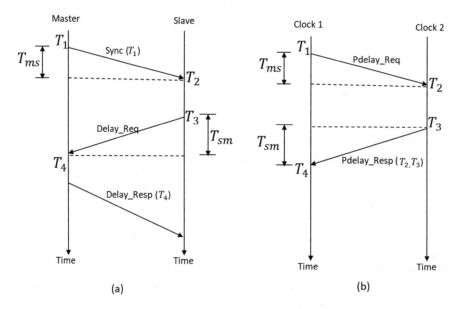

Fig. 8.5 Synchronization with delay request-response response mechanism (**a**) and synchronization with peer delay mechanism (**b**)

- T_1 = Master time at the point of sending Sync message
- T_2 = Slave time at the point of receiving Sync message
- T_3 = Slave time at the point of sending Delay_Req message
- T_4 = Master time at the point of receiving Delay_Req message

Only one of the delay mechanisms as depicted in the figure above can be used in a PTP domain for specific a period of time. The offset of slave clock from the master clock can be calculated as follows:

$$\text{Offset} = (T_2 - T_1)\_\text{Path}\_\text{Delay}$$

[where Path_Delay = $(Tms + Tsm)/2 = \dfrac{(T_4 + T_3) + (T_2 + T_1)}{2}$].

If peer delay method is in use, the path delay (excluding the last link to the slave) is carried in the correction field of the Sync message [1]. The figure above is a typical of one-step methods of synchronization where the Sync and Pdelay_Resp messages carry T_1 and (T_2, T_3), respectively. In one-step mode, fewer PTP messages are sent and it is mostly one-way communication from master to slave in which time-stamp is added to the sync message on the fly. According to IEEE1588-2019, a device port implementing one-step mode will not send follow-up message. To be more accurate, Sync or Pdelay_Resp message will not be followed by a Follow_Up or Pdelay_Resp_Follow_Up message, respectively. Conversely, two-step mode in which the Sync and Pdelay_Resp messages are followed by Follow_Up and Pdelay_Resp_Follow_Up messages carrying T_1 and T_3, respectively. Figure 8.6 depicts a simplistic purview of one-step and two-step synchronization mode.

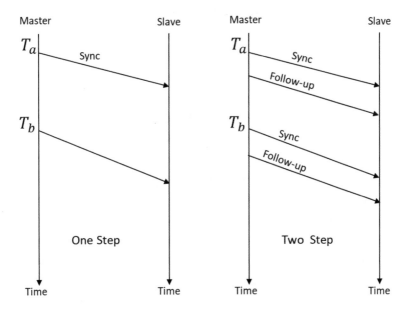

Fig. 8.6 A simplistic purview of one-step and two-step synchronization mode

One bit of caution, underlying port speed may be a factor of choosing one-step or two-step as devices 10GbE PHY may not have enough time to insert hardware timestamp, thus limiting one-step mode for 10GbE ports.

We discussed this but re-emphasizing again here: for PTP two-step operation, network must be designed in such a way that general message takes the same path (Sync and Follow_Up), or in some cases the reverse path (Delay_Req and Delay_Resp), as the event message through a transparent clock. Failure to do so will result in a condition where PTP does not calculate path delay properly. This condition is undetectable and may introduce additional jitter and wander, but it will not cause the operation of the PTP protocol to cease [2].

8.2.4   PTP Message and Packet Structure

We discussed that various PTP message classes can be divided in two categories: event and general PTP message. The event messages (Sync, Delay_req, Pdelay_Req, and Pdelay_Resp) are timestamped and generated during transmission and receipt. IEEE1588-2019 present a model on how the timestamp is generated. Figure 8.7 depicts a timestamp general model presented in IEEE1588-2019.

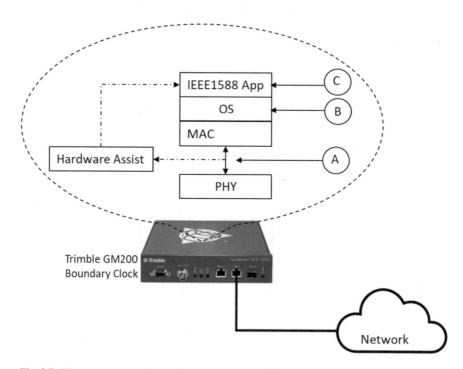

Fig. 8.7   Timestamp generation model presented in IEEE1588-2019 standard

A PTP device (here in Trimble GM 200 boundary clock as depicted above) may have multiple PTP instances implemented. For this discussion, let's assume that PTP messages are issued in PTP application code in one PT instance and received and processed by the other. These PTP messages typically have a preamble associated with underlying communication protocol (e.g., ethernet) that is used for the PTP transport. The preamble is followed by one or more protocol specific headers and then user data such as the PTP payload. When a PTP message passes through various points identified in the diagram above, a timestamp is generated in the defined point. For example, the defined point could be application layer as illustrated by "C" or OS kernel level as identified by "B" or physical technology level as identified by "A." In general, timing error will be less for the point that is closer to the network. The dotted line between PHY and MAC illustrates hardware circuitry for timestamp at physical layer.

Various PTP messages (whether event or general PTP messages) are transported over underlying technology header. IEEE1588-2019 has defined PTP transport over six different transport technologies: Ethernet, UDP in IPv4 networks, UDP in IPv6 networks, DeviceNet, ControlNet, and IEC61158 type 10. In a typical IPv4 network, PTP messages are encapsulated with UDP header as depicted below and all event messages including Sync and use UDP destination port 319. On the other hand, general messages use UDP destination port 320.

8.2.4.1   PTP Messages Over UDP in IPv4 Network

For IPv4 network, PTP messages are encapsulated with UDP header and placed in place of UDP datagram as shown in Fig. 8.8. The packet structure contains Ethernet header, IP header, UDP header and datagram and checksum fields.

The ethernet header includes EtherType field that indicates next level header as IPv4. The value of EtherType field for such packet is 0x0800. The IP header includes among other field a protocol field and destination IP address fields. The protocol field contains a value of 17 for UDP indicating type of data for the payload. The

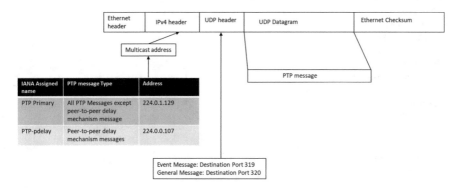

Fig. 8.8 PTP transport over UDP in IPv4 network

destination IP address should contain one of the two multicast address: 224.0.1.129 and 224.0.0.107. As shown in the figure above, 224.0.1.129 should be used for all PTP messages except peer-to-peer delay mechanism messages. For the later, 224.0.0.107 multicast address is used as the IP destination address. The next level header is UDP that contains among other fields a destination port field. The destination UDP port for event messages is 319 while general message should use destination UDP port 320. All PTP messages are part of UDP datagram in a IPv4 network.

8.2.4.2 PTP Messages Over UDP in IPv6 Network

For IPv6 network, UDP port remains the same but multicast addresses changed as depicted in Table 8.2.

For PTP messages, the Differentiated Service (DS) field in the Traffic Class (TC) field of IPv6 header must be set to highest traffic class selector codepoint. Please refer to RFC 2460 for IPv6 TC field.

Table 8.2 IPv6 multicast address for PTP messages [2]

IANA assigned name	PTP message types	Address (hex)
PTP-primary	All PPTP messages except peer-to-peer delay mechanism messages	FF0X:0:0:0:0:0:0:181 (please see notes)
PTP-pdelay	Peer-to-peer delay mechanism messages	FF02:0:0:0:0:0:0:6B

Notes
Section 2.7 of RFC 4291 (IPv6 addressing structure) defines hexadecimal values for "X" in the PTP-primary address as follows:
0 Reserved
1 Interface-local scope
2 Link-local scope
3 reserved
4 Admin-local scope
5 Site-local scope
6 (unassigned)
7 (unassigned)
8 Organization-local scope
9 (unassigned)
A (unassigned)
B (unassigned)
C (unassigned)
D (unassigned)
E Global scope
F reserved

8.2.4.3   PTP Messages Over Ethernet

For Ethernet, PTP messages are directly carried in the ethernet payload after the header as shown in the figure below. Ethernet header contains among others a destination and a type/length field. IEEE1588-2019 specifies multicast MAC address of 01-1B-19-00-00-00 for all PTP messages except peer-to-peer delay mechanism messages. The later uses 01-80-C2-00-00-0E as the multicast MAC address for the destination MAC address of ethernet header. According to IEEE1588-2019, it is permissible to use address 01-1B-19-00-00-00 or address 01-80-C2-00-00-0E for all PTP messages if such use is defined in the applicable PTP Profile (Fig. 8.9).

It is to be noted that as per IEEE Std 802.1Q-2014, frames containing 01-80-C2-00-00-0E in their destination address field are not relayed by the bridge. Therefore, this address is selected for the peer-to-peer delay mechanism messages because the scope of this address is limited to an individual LAN, and this is the normal case of the PTP peer-to-peer delay mechanism messages. This address is not assigned exclusively to PTP, but rather it is a shared address [2].

The Type field of Ethernet header should be set to an ethertype value of 88F7 as shown in the figure above indicating data type in the payload. PTP messages are inserted in the payload section of ethernet packet. PTP packet should always have highest traffic class for which a definition and implementation of Traffic class is specified in IEEE Std 802.1Q-2014.

8.2.4.4   PTP Messages Structure

All PTP messages carried over underlying technologies (e.g., ethernet and UDP) have header, body, and suffix as shown in Fig. 8.10. The common PTP header is 34 bytes in length, body is set to variable length while suffix is optional and set to 0 length.

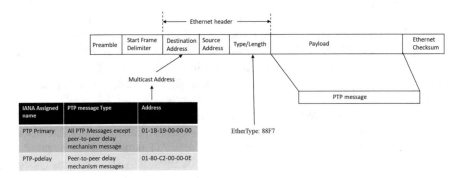

Fig. 8.9 PTP messages over Ethernet transport [2]

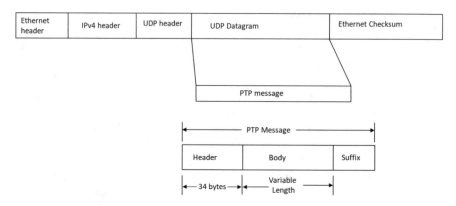

Fig. 8.10 PTP message structure

Bits								Octets	Offset
7	6	5	4	3	2	1	0		
majorSdoId				messageType				1	0
minorVersionPTP				versionPTP				1	1
messageLength								2	2
domainNumber								1	4
minorSdoId								1	5
flagField								2	6
correctionField								8	8
messageTypeSpecific								4	16
sourcePortIdentity								10	20
sequenceId								2	30
controlField								1	32
logMessageInterval								1	33

Fig. 8.11 PTP message header as defined in IEEE1588-2019

The common PTP message header in PTP version 2.1 defined by IEEE1588-2019 includes some fields that are not present in version 2.0 defined by IEEE1588-2008. Figure 8.11 illustrates PTP message header as defined by IEEE1588-2019.

The majorSdoId field, a 4 bits field was transportSpecific field in IEEE1588-2008 or PTP version 2. The transportSpecific field in PTP version 2 distinguishes the IEEE 802.1AS profile from all other PTP profiles. Figure 8.12 depicts a PTP packet capture indicating transportSpecific field in PTP version 2.

This field along with minorVersionPTP and minorSdoID field collectively create *PTP profile isolation*. So, what is PTP profile isolation? Let's consider, if two or more profiles use multicast in the same network, then PTP nodes will see the profile which are of no interest. This will cause havoc with BMCA (Best Master Clock Algorithm). In PTP version 2, the way to overcome this is to use a PTP domain for

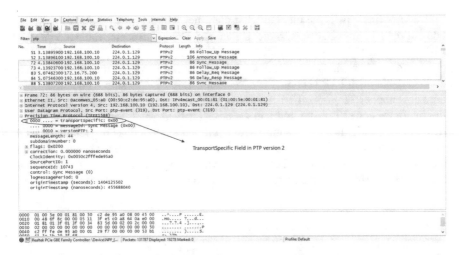

Fig. 8.12 PTP version 2 packet capture depicting transportSpecific field

Table 8.3 Various message class and their values in messageType field

PTP message type	Message class	Value (hex)
Sync	Event	0
Delay_Req	Event	1
Pdelay_Req	Event	2
Pdelay_Resp	Event	3
Reserved	–	4–7
Follow_Up	General	8
Delay_Resp	General	9
Pdelay_Resp_Follow_Up	General	A
Announce	General	B
Signaling	General	C
Management	General	D
Reserved	–	E–F

each profile. This works for network administrator who are careful in the PTP network design and manually configure domain-specific PTP profiles. But as such is not viable for many who may not be aware of this issue or may not have experience to configure this. Certain standard development organization (SDO) also like to have their unique identifier such as SdoId that appears in the common header of all PTP messages and PTP nodes can ignore all PTP messages that do not have SdoId they want. An organization can get this SdoId from IEEE Registration Authority. The idea here is that a SDO can get their own SdoId and protect its own various profiles from each other using rules on domains. It is to be noted that only standard body will get the SdoId but not any other organization. The SdoId in PTP version 2.1 header comprises of two fields: majorSdoId (4 bits) and minorSdoId (8bits).

The messageType is a 4 bits field that identifies which type of message is in the body of the PTP message, e.g., Sync, Delay_req, Delay_resp, etc. Table 8.3 depicts various message types and corresponding value in hex.

The minorVersionPTP and versionPTP fields are 4 bits each that allows a PTP node to distinguish PTP version 2 and Version 2.1. A PTP Version 2 node that does not know about PTP Version 2.1 receives a PTP Version 2.1 message, it will ignore the minorVersionPTP value and use versionPTP field value for processing. Next field in PTP header is a 2 bytes messageLength field that defines full length of PTP message. The domainNumber field is a 1 byte field that identifies specific domain the PTP message belongs to. The PTP domain is logical grouping of clocks that synchronizes with each other using PTP, they may not synchronize with other

Table 8.4 The values of flagField as defined in IEEE1588-2019

Octet	Bit	Message types	Name	Description
0	0	Announce, Sync, Follow_Up, Delay_Resp	alternateMasterFlag	FALSE if the PTP Port of the originator is in the MASTER state, otherwise set to TRUE
0	1	Sync, Pdelay_Resp	twoStepFlag	For Sync messages, if there is a Follow_Up message associated with the Sync message the twoStepFlag to TRUE, otherwise FALSE For Pdelay_Resp messages, if there is a Pdelay_Resp_Follow_Up message associated with the Pdelay_Resp message the twoStepFlag to TRUE, otherwise FALSE
0	2	ALL	unicastFlag	TRUE, if PTP message was sent to unicast address. FALSE, if PTP message was sent to multicast address
0	5	ALL	PTP profile specific 1	Defined by PTP profile; otherwise FALSE
0	6	ALL	PTP profile specific 2	Defined by PTP profile; otherwise FALSE
0	7			Reserved
1	0	Announce	leap61	The value of timePropertiesDS.leap61
1	1	Announce	Leap59	The value of timePropertiesDS.leap59
1	2	Announce	currentUtcOffsetValid	The value of timePropertiesDS. currentUtcOffsetValid
1	3	Announce	ptpTimescale	The value of timePropertiesDS. ptpTimescale
1	4	Announce	timeTraceable	The value of timePropertiesDS. timeTraceable
1	5	Announce	frequencyTraceable	The value of timePropertiesDS. frequencyTraceable
1	6	Announce	synchronizationUncertain (optional flag)	Optional field

domain clocks. The flagField contains various flags to indicate status. It is a 2 octets field. Table 8.4 depicts values of the flagField as defined in IEEE1588-2019.

The next field is correctionField that contains correction value in nanosecond for residence time within a transparent clock and path delay for peer-to-peer transparent clocks. The messageTypeSpecific is a 4 bytes field that is used for internal synchronization of PTP instance and its ports. For example, if the PTP Instance consists of multiple hardware components such as SOC and processor that are not synchronized, messageTypeSpecific can be used to transfer an internal timestamp between these components.

The sourcePortIdentity is a 10 bytes field that identifies originating port for the PTP message. The sequenceId is next field that contains sequence number for individual message type. Next is controlField, a 1-byte field. For all traffic except IPv4 transport, this field is obsolete and should be ignored. This field is used for hardware compatibility during PTP over IPv4/UDP transport. When supporting the hardware compatibility, the value of controlField depends on the message type defined in the messageType field, and it shall have the value specified in Table 8.5.

Next is logMessageInterval, a one-byte field whose value is determined by type of PTP message.

8.2.4.5 Body of the PTP Header

The body of the PTP message contains different PTP message types and respective message contents as discussed earlier. For example, if the message is Announce it should include contents that capabilities of a clock to other clocks within a domain. This allows master-slave hierarchy to be established.

Announce Message

Figure 8.13 shows Announce message format that will be inserted in the body section of PTP header.

The header section is a 34 bytes field that includes header as discussed earlier. The next field is originTimestamp, a 10-bytes field whose value is generally shall be

Table 8.5 controlField parameters

Message type	controlField value (hex)
Sync	00
Delay_Req	01
Follow_up	02
Delay_Resp	03
Management	04
All other	05
Reserved	06 to FF

Bits								Octets	Offset
7	6	5	4	3	2	1	0		
Header								34	0
originTimestamp								10	34
currentUtcOffset								2	44
reserved								1	46
grandmasterPriority1								1	47
grandmasterClockQuality								4	48
grandmasterPriority2								1	52
grandmasterIdentity								8	53
stepsRemoved								2	61
timeSource								1	63

Fig. 8.13 Announce message format [2]

Bits								Octets	Offset
7	6	5	4	3	2	1	0		
Header								34	0
originTimestamp								10	34

Fig. 8.14 Sync and Delay_Req message format [2]

0 or an estimate no worse than ±1 s of the PTP Instance Time of the originating PTP Instance when the Announce message was transmitted. PTP provides UTC offset and Leap second to convert time between TAI and UTC in the end node and this value is carried in the currentUtcOffset. The grandmasterPriority1 field indicates the priority value of the grandmaster clock. Lower value takes precedence. The grandmasterClockQuality field value is based on attributes in the BMCA. Next is grandmasterPriority2 field that indicates the priority value of the grandmaster that is used when Grandmaster Priority1 value is the same for different masters in a network. The grandmasterIdentity is the clock identity of the grandmaster. It is a 64-bit global identifier (EUI-64) as defined by the IEEE standard. The stepsRemoved field indicate the number of boundary clocks between the local clock and the foreign master clock. The timeSource field indicates the nature of the source of time and frequency distributed by the Grandmaster PTP Instance.

Sync and Delay_Req Message

The sync request is sent by master clock and contains the master time. If the Master is a two-step clock, timestamp in the sync message will be set to 0 and actual sending timestamp will be sent afterwards in the associated with Follow_Up message. The Sync message is sent in both types of delay measurement mechanisms. Both

Sync and Delay_req messages have similar message format. Figure 8.14 shows a typical Sync and Delay_Req message.

The Delay_Req message is sent by a slave clock and contains slave time. This message is sent only in delay request-response mechanism.

Follow_Up Message

The Follow_Up message is optionally sent by a master clock and contains master clock when the sync message was sent. It is used by two-step master clock (Fig. 8.15).

The value of the preciseOriginTimestamp field of the Follow_Up message shall be an estimate no worse than ± 1 s of the timestamp of the associated Sync message excluding any fractional nanoseconds.

Delay_Resp Message

The Delay_resp message is sent by master clock and contains master time when the Delay_req message was received (Fig. 8.16).

The receiveTimestamp field of the Delay_Resp shall be the timestamp of the associated Delay_Req message. The requestingPortIdentity is the source port identity associated with the header of Delay_req message.

Pdelay_Req Message

The Pdelay_Req is sent by a "delay requester" peer-to-peer clock and contains "delay requester" peer-to-peer clock time when the pdelay_req message was sent. This message is only sent in the peer delay mechanism. The following diagram shows Pdelay_resp message format (Fig. 8.17).

The reserved field in the Pdelay_Req message is to make the PTP message length match the length of the Pdelay_Resp message. In some networks and bridges, PTP messages with unequal lengths have different transit times that introduce asymmetry errors [2].

Bits								Octets	Offset
7	6	5	4	3	2	1	0		
Header								34	0
preciseOriginTimestamp								10	34

Fig. 8.15 Follow_Up message format [2]

Bits								Octets	Offset
7	6	5	4	3	2	1	0		
Header								34	0
receiveTimestamp								10	34
requestingPortIdentity								10	44

Fig. 8.16 Delay_Resp message format [2]

Bits								Octets	Offset
7	6	5	4	3	2	1	0		
Header								34	0
requestReceiptTimestamp								10	34
requestingPortIdentity								10	44

Fig. 8.18 Pdelay_Resp message format [2]

Bits								Octets	Offset
7	6	5	4	3	2	1	0		
Header								34	0
origineTimestamp								10	34
reserved								10	44

Fig. 8.17 Pdelay_Resp message format [2]

Pdelay_Resp Message

The Pdelay_Resp message is sent by a "delay responder" peer-to-peer clock and contains "delay responder" peer-to-peer clock time when Pdelay_Req message was received. It is sent only in peer delay mechanism. Figure 8.18 depicts Pdelay_Resp message format.

Pdelay_Resp_Follow_Up Message

The Pdelay_Resp_Follow_Up message is optionally sent by a "delay responder" peer-to-peer clock and contains "delay responder" peer-to-peer clock time when the Pdelay_Resp was sent.

The message format depicted in Fig. 8.19 is used when the "delay responder" is a two-step clock. It also has option to send a turnaround time instead of sending timestamp.

Bits								Octets	Offset
7	6	5	4	3	2	1	0		
Header								34	0
responseOrigineTimestamp								10	34
requestingPortIdentity								10	44

Fig. 8.19 Pdelay_Resp_Follow_Up message format [2]

Bits								Octets	Offset
7	6	5	4	3	2	1	0		
Header								34	0
targetPortIdentifier								10	34
One or More TLV								M	44

Fig. 8.20 Signaling message format [2]

Signaling Message

The signaling message are used by PTP clocks to negotiate certain optional services, such as unicast transmission. PTP uses multicast communication by default. The message includes targetPortIdentifier and one or more TLV (Type, Length, Value). The targetPortIdentifier field contains the address of target port or ports of the message. It is 10 bytes field.

Figure 8.20 depicts "Signaling" message format.

Management Message

The PTP Management message is used to transmit information from a clock to a node manager and from a node manager to one or more clocks. The targetPortIdentity is a 10 bytes field that contains the address of port or ports of the message (Fig. 8.21).

Next is startingBoundaryHops, a 1-byte field contains information about number of boundary clock that this message is allowed to be retransmitted by. The boundaryHops field contains the number of boundary clock retransmission left for this management message.

The actionField contains type of action that this management message is required to perform.

Bits								Octets	Offset
7	6	5	4	3	2	1	0		
Header								34	0
targetPortIdentity								10	34
startingBoundaryHops								1	44
boundaryHops								1	45
reserved		actionField						1	46
reserved								1	47
managementTLV								M	48

Fig. 8.21 PTP Management message format [2]

Bits								Octets	Offset
7	6	5	4	3	2	1	0		
tlvType								2	0
lengthField								2	2
managementId								2	4
dataField								N	6

Fig. 8.22 Management TLV format [2]

Management TLV

The management TLVs mentioned in PTP management message are explained in this section. The management TLV includes tlvtpe, lengthField, managementId, and dataField as shown in the figure below.

The tlvType should be set to MANAGEMENT (0x0001). The lengthField is the length of TLV for which the format is $2 + N$ where N is even number (Fig. 8.22).

The managementID field defines type of management message, e.g., Initialize, Enable_Port, and Disable_Port.

8.2.5   PTP Integrated Security Mechanism

This is an optional feature and supported in PTP version 2.1 defined by IEEE1588-2019 as security extension. The security extension specifies authentication, message integrity, and replay attack protection for PTP messages within a domain. Additionally, PTP version 2.1 also provides support for external security mechanisms such as MACSEC and IPSEC. In this section, we will discuss these internal and external security support of PTP messages.

8.2.5.1   Authentication TLV

The security extension specified in IEEE1588-2019 is implemented by following two mechanisms:

1. *Authentication TLV*: The authentication TLV is used by grandmaster or master clock to the slave. The message carries all necessary information to enable security processing for the sender and the receiver. PTP version 2.1 allows multiple authentication methods to be performed using symmetric keys. Depending upon associated key management, authentication TLV supports two different verification methods: immediate security processing and delayed security processing. Immediate security processing involves the processing of the authentication TLV before the content of the PTP message is processed. This is performed by a key management that allows shared secrets for the pair (as in unicast communication) or for the group (as in multicast communication). On the other hand, delayed security processing involves delayed distribution of security parameters, e.g., shared secrets. It requires the storage of messages on the receiver side until the required security parameters have been provided for a posteriori verification of the stored PTP messages.
2. *Associated Key Management*: This method allows for the secure distribution of all necessary security parameters required to construct or verify the authentication TLV. User may choose their own implementation of key management and hence, it can be manual or automated. There are few industry-defined key management mechanisms, e.g., GDOI (Group Domain of Interpretation) for immediate security verification and TESLA (Timed Efficient Stream Loss-Tolerant Authentication) for delayed security verification.

Figure 8.23 depicts PTP security extension message format. For each authentication TLV processing mechanism an ICV (Integrity Check Value) field is inserted.

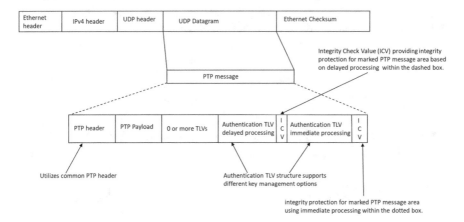

Fig. 8.23 Authentication TLV message format [2]

PTP version 2.1 allows multiple authentication TLVs if the PTP message header field which is protected by authentication TLV remains unchanged. Since PTP security extension also ensures the integrity of PTP instances, it is required that key management verifies the authenticity of PTP instances.

For the calculation of ICV, the key management must provide the following information:

- Security Parameter Point (SSP): Identifies the Security Association (SA) negotiated by the key management protocol.
- IntegrityAlgTyp: Identifies integrity algorithm type used to compute the ICV associated with each SPP.
- icvLength: Indicates the length of the calculated ICV.
- Key: A symmetric key to be used in conjunction with the selected algorithm.
- keyLength: Indicates the length of the disclosed key.
- sequenceNo: Indication for use of sequenceNo and the desired length of the sequenceNo field.
- sequenceID: Indication of the sequenceID window for anti-replay.
- Immediate Security: A Boolean indicating whether the SA is for immediate security or delayed security.
- RES: Indication for use of RES and the desired length of the RES field (please refer to figure below).

Figure 8.24 shows various fields of authentication TLV.

The values of the authentication field are determined by the message format provided below. The tlvType is a 2 octets field for which the value should be authentication. Next is lengthField that indicates the length of authentication TLV payload. The value of lengthField is $6 + D + S + R + K$. Please note D, S, R, and K shall be even values, and each value shall represent the number of octets in the field. SPP value should match with those security parameters provided by the key management. Next is the secParamIndicator, a 1-byte field. It is optional and indicates presence or absence of optional field. The value of KeyID field depends on whether the implementation is immediate security processing or delayed security processing. All other fields (disclosedKey, sequenceNo, and RES) are optional except ICV. The

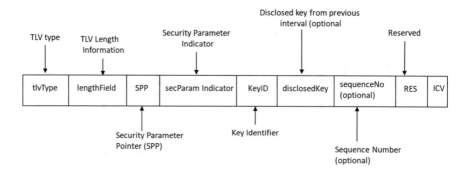

Fig. 8.24 Various Authentication TLV fields [2]

Bits								Octets	Offset
7	6	5	4	3	2	1	0		
tlvType								2	0
lengthField								2	2
SPP								1	4
secParamIndicator								1	5
keyID								4	6
disclosedKey (optional)								D	10
sequenceNo (optional)								S	10 + D
RES (optional)								R	10+D+S
ICV								K	10+D+S+R

Fig. 8.25 Authentication TLV fields and values [2]

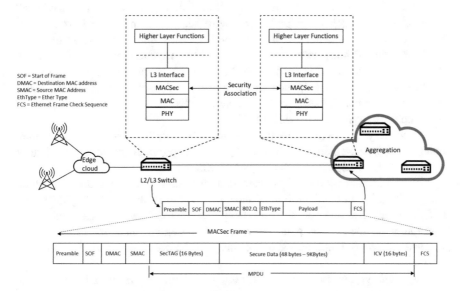

Fig. 8.26 A typical network with MACsec implementation [2]

ICV shall have a data type of UInteger with a length of K octets. The value of K depends on the algorithm as discussed below but must be an even integer.

This field shall contain the calculated ICV of the PTP message carrying the authentication TLV. The ICV is calculated according to the IntegrityAlgTyp provided by the key management (Fig. 8.25).

The implement should at least have HMAC-SHA256-128 defined in FIPS PUB 198-1. User interested to learn more please refer to IEEE1588-2019.

8.2.5.2   PTP Over MACsec

The MACsec is a layer 2 security protocol that is implemented in ethernet hardware. It is an extension to the 802.1X and provides secure key exchange and mutual authentication for MACsec nodes. PTP version 2.1 specifies mechanism to run PTP over MACsec in a layer 2 network. It was standardized in 2006 by IEEE (standard IEEE 802.1AE-2006) which requires a MACSec Key Agreement (MKA) that is defined by IEEE 802.1X-2010. The MKA provides key exchange and allows mutual authentication of nodes that want to take part in a MACsec security association. As shown in Fig. 8.26, MACsec is used for secured association between a L2/L3 switch at the edge cloud and L2/L3 switch at the metro aggregation. This setup allows secured flow of traffic from edge cloud to the aggregation. Since MACsec is point to point level, it creates a secure traffic flow between the endpoint that implements MACsec, herein between the edge switch and aggregation switch. The implementation of MACsec supports both layer 2 and layer 3 connectivity. A PHY/MAC SOC (system on chip) generally implements MACsec and from layered concept, its implementation is between L3 interface and MAC (Medium Access Control).

On the wire, frames are encapsulated with MACsec encryption as shown in the figure below. The SecTAG field in the MACsec frame carries information about established secure channel, such as Ethertype, TAG control information, secure association number, packet number, and secure channel identifier. The SecTAG is followed by encrypted payload and ICV.

Collectively, SecTAG, secure payload and ICV is called MACsec protocol data unit (MPDU). Any ethernet payload irrespective L3 protocol and datagrams thereof will be encrypted as secure data and part of MPDU.

The MACsec treats the PTP packet same way as it does any ethernet payload irrespective of protocol definition as shown in the figure above. Ideally, timestamp is done in physical layer but the timestamp measurement can be performed before decryption on the ingress path. However, on the egress path, the timestamp or updated correctionField needs to be inserted before encryption, for one-step operation.

Since encryption/decryption can introduce high delays in PTP messages, IEEE1588-2019 suggests the following two methods for high accuracy:

- If the encryption/decryption module is implemented with a low delay variation, timestamping can be performed accurately at the PTP layer labeled "IEEE1588" (Please refer to figure below).
- In two-step mode, timestamping can be performed at the physical layer since the transmitted PTP messages do not require in-flight modification.

Figure 8.27 shows recommended timestamping before and after MACsec functions for ingress and egress path as discussed earlier.

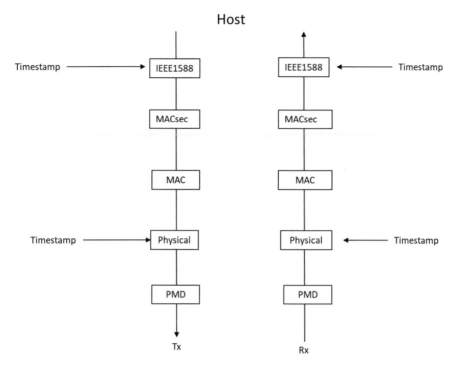

Fig. 8.27 Timestamping before and after MACsec functions as recommended by IEEE1588-2019

8.2.5.3 PTP Over IPSec

Defined by a series of RFCs (RFC1825, 1826, and 1827), Internet Protocol Security (IPSec) is a tunneling protocol suite that provides secure encrypted communication between two endpoints over an Internet Protocol (IP) network. It is normally used for VPN (Virtual Private Network) solutions. Since PTP is used over IP network, IPSec tunneling become useful for securing IP layer payload including PTP packets as UDP datagrams. PTP version 2.1 refers to following RFCs for IPSec operation and those are: RFC 7619, RFC8247, and RFC 4303. These RFCs collectively provide authentication, key management and integrity for Encapsulating Security Payload (ESP) tunnel mode. For the PTP operation using IPSec, we could consider two network environments: (1) PTP tunneled over non-PTP aware network and (2) PTP in a PTP-aware network.

For the first scenario, let's consider grandmaster clock located in trusted network and an ordinary clock or boundary clock located in a public network as depicted in Fig. 8.28. The grandmaster Trimble's GM200 is in a data center within a trusted network and connected through a IPSec secure gateway to retail network endpoint over a public network. Such scenario will be common in future CBRS-based network. We will explore this scenario in the proceeding chapter. Assuming that public network can guarantee and prioritize PTP traffic, CBRS radio unit can provide data

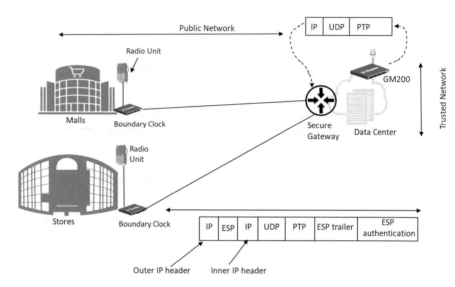

Fig. 8.28 Retail CBRS deployment using IPSec tunnel for time synchronization

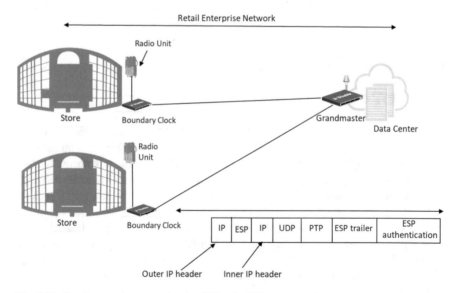

Fig. 8.29 Retail enterprise network using IPSec for PTP

and mobile services to retail endpoints where it is mall or wholesale stores like Costco.

In this time synchronization service, IPSec tunnel mode should be using ESP Tunnel mode as specified in IETF RFC 4303 and Internet Key Exchange (IKE) v2. The IPSec tunnel will be established between far end boundary clock and IPSec secure gateway. From PTP perspective IEEE1588-2019 suggests that any encryption algorithm can be used or the traffic can be without encryption. However, PTP traffic must be guaranteed with highest priority using DSCP (Differentiated Service Code Point). Therefore, mapping of DSCP between outer IP header and inner IP header is needed.

In case of a PTP-aware network where there is no IPSec secure gateway, grand-master clock and boundary clock will implement adequate IPSec services. Let's consider that a retail enterprise like Costco has many stores in a region for which they would like to deploy CBRS network for data and mobile services. In this case, an endpoint IPSec tunnel is established between grandmaster and the boundary clock. Time synchronization is thus performed through PTP packets that are encapsulated as IPSEC payload and exchanged between grandmaster and boundary clock (Fig. 8.29).

However, please note that IPSec brings some degradation of the timestamping accuracy due to the additional security processing if the attached PTP Node and PTP Instances are not designed properly. If the IPSec is implemented in hardware, then same principle applies as discussed in the MACsec section. In case of software implementation of IPSec, the following two approaches should be considered:

1. Depending on the timing requirements, it might be sufficient to use software-based timestamping, potentially causing lower accuracy. In the case of CBRS deployment, as discussed earlier, if the deployment is for data services then perhaps software implementation of IPSec can be accepted assuming that it does not introduce more than few microseconds in delay.

2. Hardware-based timestamping in combination with IPsec encryption requires some further considerations, as PTP messages are sent over a software-based IPsec tunnel. On the ingress path, it is not possible to identify incoming PTP messages until after decryption. On the egress path, it is not possible to update the timestamp field in an outgoing PTP message after it has been encrypted. Therefore, all incoming PTP messages have to be timestamped and stored until the corresponding PTP message is decrypted and identified as a PTP message. For outgoing PTP event messages it is recommended to use two-step mode, thus allowing transmission of encrypted PTP messages that contain accurate time-stamps. The outgoing PTP event messages needs to be identified in the hardware by an implementation specific method, for example by an internal flag attached to the message[2]. For further details about design related to software-based IPSec implementation, please consult IEEE1588-2019.

Additionally, IEEE1588-2019 recommends redundancy design for timing implementation where delay attacks cannot be thwarted by cryptographic protection.

8.3 PTP Profile

PTP defines a number of attribute options (such as PTP over ethernet and PTP over IP/UDP) and features, but one of the goals of PTP standard is to make it possible to set up and run a PTP network with minimal device settings and administration. This is done through PTP profiles which allows SDO or industry groups to specify a subset of options and features as well as default values for protocol attributes that will meet the performance requirements of applications in the domain and eliminate or minimize device settings. It can extend the protocol by defining an ABMCA, providing new transport mapping, and appending profile-specific tag, length, value (TLV) triplets to PTP messages. There are a number of PTP profiles being defined in various standards, for example default profile (two basic general purpose profiles) is defined by IEEE1588-2008 standard whereas telecom profiles are defined by ITU-T G.8265.1, G.8275.1, and G.8275.2, enterprise profile is defined by an IETF draft (https://tools.ietf.org/html/draft-ietf-tictoc-ptp-enterprise-profile-09), power profile is defined by IEC 61850-9-3, IEEE C37.238-2011 and C37.238-2014, and PTP media profile defined by SMPTE ST-2059-2, AES67 and AES67 + SMPTE ST-2059-2.

8.3.1 Default Profile

The PTP standard defines the two default profiles which are identical except for the method of delay measurement specified: one for the delay request-response mechanism and the other for the peer delay mechanism. These two profiles specify the default values for the protocol attributes, but other attributes and features are still optional. Both profiles use Best Master Clock Algorithm (BMCA) to determine which node on the network is master. It is not unusual for a network to consist of nodes using either of the default protocols. Although default profile specifies certain parameters, network administrators may be able to configure some of these.

The delay request-response default PTP profile is used for end-to-end delay measurement. In such configuration, slaves send periodic delay requests over multicast for which a master receiving the request sends a multicast delay response packet with the slave request's sequence number and portIdentity. All nodes on the network can see each request and response, even though delay requests are only answered by masters, and a particular delay response is only meaningful to the slave listening for that particular packet. The delay request frequency is controlled by master for which interval is communicated in regular message.

The Peer-to-Peer Default PTP Profile is used for peer-to-peer delay measurement. The operation is similar to end-to-end delay measurement, but primarily used by boundary clocks, or between multiple appliance-type grandmasters, rather than by individual nodes.

8.3.2 Telecom Profiles

There are three standards that collectively define telecom profiles, and these are: ITU-T G. G.8265.1 (frequency Profile), ITU-T G.8275.1 (Phase and Time Profile), and ITU-T G.8275.2 (Partial timing and Assisted Partial Timing). We will discuss telecom profiles in details in the proceeding chapter.

8.3.3 Enterprise Profile

Although this profile is still under definition by a IETF draft (https://tools.ietf.org/html/draft-ietf-tictoc-ptp-enterprise-profile-09), the profile is widely used. It is a specialized version of the default end-to-end profile, and is interoperable with the default end-to-end profile. It provides potentially tighter synchronization, allows for slaves to follow masters in multiple domains, and eliminates the need to query the masters for network variables.

8.3.4 Power Profile

As series of standards defines power profiles and those are: IEC 61850-9-3, IEEE C37.238-2011, C37.238-2014, C37.238-2017 and IEC/IEEE 61850-9-32,016. We will discuss these profiles in the proceeding chapter related to Smart Grid infrastructure.

8.3.5 PTP Media Profile

PTP media profile comprises of three profiles: SMPTE ST-2059-2, AES67, and AES67 + SMPTE ST-2059-2. These profiles support various audio and video applications for capture to be used in professional broadcast environments. The SMPTE ST-2059-2 standard defines SMPTE profile for time and frequency synchronization in professional broadcast environment. It allows multiple video sources to stay in synchronization across various equipment by providing time and frequency synchronization to all devices. This standard is intended to be used with SMPTE ST 2059-1 which defines a point in time (the SMPTE Epoch) used for aligning real-time signals and formula for ongoing signal alignment [3]. The profile is defined with following objectives:

- Provide the default values of configurable attributes that allows a slave to be synchronized within 5 s of its connection to the operational PTP network.
- After initial synchronization, slave device should maintain time accuracy with respect to the master reference within 1 μs.
- To convey Synchronization Metadata (SM) required for synchronization and time labeling of audio/video signals.

The AES-67 profile is based on AES67 standard and used for high quality audio synchronization over IPv4 multicast networks. This profile enables audio streams to be combined at a receiver and maintain stream synchronization.

The AES67+ SMPTE ST-2059-2 combines both AES-67 and SMPTE ST-2059-2 for audio and video synchronization. This profile enables two standards to work together over the same network.

References

1. Watt, T. S., Achanta, S., Abubakari, H., & Sagen, E. (2015). *Understanding and applying precision time protocol*. Schweitzer Engineering Laboratories, Inc.
2. IEEE-1588. (2019). *IEEE standard for a precision clock synchronization protocol for networked measurement and control systems*. IEEE Instrumentation and Measurement Society.
3. SMPTE ST 2059-2. (2015). *SMPTE STANDARD: SMPTE profile for use of IEEE-1588 precision time protocol in professional broadcast applications*. SMTPE. The Society of Motion Picture and Television Engineers.

Chapter 9
Synchronization for Telecom Infrastructure

9.1 Introduction

Over the last few decades telecom infrastructure undergone tremendous changes from analog to digital circuit oriented to virtualized, on demand and service-oriented infrastructure. The changes that occurred in telecom infrastructure are influenced by two distinct technologies: first, wireless or RAN technologies and the second is network technologies. From historical perspective each of these technologies enabled other to realize the best services that could be rendered in the telecom infrastructure for a given period of time. For example, technologies such as PDH that transformed network systems as digital interconnect also enabled end-to-end transport of digital mobile communications in 2G and so on. While 1G to 4G mobile network transformation was considered leapfrogged, 5G is a significant leap forward. It changed the nature of mobile network infrastructure due to its innate separation of RAN and core network. More importantly, 5G allowed whooping bandwidth capabilities at the wireless endpoints and capability to support demanding traffic. However, 5G would not alone have changed the telecom infrastructure to become on demand and service-oriented if not open networking and other technology catalysts worked collectively to facilitate network decomposition. As telecom infrastructure renders in telecom cloud, synchronization becomes both difficult and a must for the future of network. In this chapter, we begin with historical purview of technological changes that occurred at both mobile network and core of telecom networks and how they were linked. As we discussed the progression and transformation of telecom network from the 1G era to 5G era, synchronization for each type of network is discussed. More importantly, the link between mobile network transformation and network transformation as it relates to telecom network infrastructure is discussed.

Readership will find this chapter useful in understanding technology trends in networking, mobile communications and telecom transport and how these

© The Author(s), under exclusive license to Springer Nature
Switzerland AG 2021
D. D. Chowdhury, *NextGen Network Synchronization*,
https://doi.org/10.1007/978-3-030-71179-5_9

technological advents are linked together. Today, collective innovation of various industry forums is working together bringing faster and more technological advents that otherwise would have been impossible to enable future of networks including telecom cloud.

9.2 Historical Perspective: Synchronization in Telecom Networks

To begin our discussion on synchronization for telecom networks, a good starting point is the mobile network technologies which have seen phenomenal transformation in just few decades. From historical perspective, the mobile network era starts in 1980 with introduction of the first-generation mobile network technology or 1G. The most important innovations of 1G era includes the cellular structure, frequency reuse concept, and cell planning strategies. The 1G technology was a basic analog system designed for voice communications with a data rate of up to 2.4 kbps, which uses frequency modulation (FM), frequency division multiple access (FDMA) transmission technology, and a bandwidth of 30 kHz [1]. In the 1G era, carrier phase synchronization [2] was studied to achieve coherent demodulation in wide-band FM communication and to suppress adjacent channel interference [3]. The second generation (2G) mobile systems were introduced in later 1980s. The transformation of mobile network from 1G to 2G was facilitated by the evolution of the information format from analog to digital modulations. It supported low bit rate data services as well the traditional speech services. Compared to 1G, the second generation (2G) systems used digital multiple access technology such as TDMA (time division multiple access) and CDMA (code division multiple access). The representative system of 2G such as Global System for Mobile Communications (GSM) employed TDMA, time division duplex (TDD), Gaussian minimum-shift keying, and 1.5 MHz per link to support better voice communication. This transformation from analog to digital systems in 2G era influenced the need for sync standardization. As mobile network system has undergone transformation from analog to digital system, so do telecom network infrastructure. The analog transmission lines which used trunk lines between exchanges to carry multiple voice channels simultaneously using FDM made up nearly all telephone systems up until the early 1960s when PDH (Plesiochronous Digital Hierarchy) was introduced.

9.2.1 PDH Network Synchronization

The Plesiochronous Digital Hierarchy (PDH) was born out of the telcos' desire to better use their cable facilities and to enhance the quality of calls. The PDH allowed the digitization of voice transmission while aggregating voice channels and better use of cables. The digitization of an analog voice channel into a 64-kbps digital channel using pulse code modulation (PCM) made it possible to use time division multiplexing (TDM) to multiplex a number of voice channels onto a trunk line known as T-carrier in North America and Japan. The European equivalent known as E-Carrier was used in Europe and rest of the world. Figure 9.1 shows typical construct of PDH network.

Prior to PDH, analog switches did not require synchronization, but the digital switching introduced in the PDH network did require synchronization (please refer to diagram above). The 64 Kbit/S digital switch/cross-connect are identified in the diagram as part of network synchronization component but not the PDH transport.

The PDH transport network is plesiochronous and does not require synchronization. Thus, the PDH network is sometimes referred to as asynchronous. The classical mechanism for passing sync in the PDH network was to use timeslot zero. In this case, the sync (clock information) is carried within timeslot zero of respective signal such as E1 or T1 and effectively recovered by the application from the incoming traffic link. The network equipment performed clock recovery from timeslot zero and used it to flywheel a local oscillator to provide sync for the applications. PDH was an integrated digital network meaning, it was capable of carrying different traffic as long as these traffics presented in digital manner.

9.2.1.1   2G Mobile Networks Synchronization

In the 2G mobile network that uses PDH as the backbone for transport, GSM radio sub-system required 0.05 ppm frequency accuracy for the radio interface [4, 5]. In order to fulfill the frequency accuracy requirements, a network designer may have considered either using GPS or rubidium-based clock input directly to BSC (Base Station Controller) or more commonly, distribute frequency from PRC through a synchronization chain. We have discussed about frequency distribution and PRC synchronization chain in Chaps. 3 and 6. Figure 9.2 shows typical GSM network with frequency synchronization. A common practice is to install PRC and synchronization distribution elements in MSC (mobile switching center) servers, and base station controller (BSC) sites to directly synchronize all transmission equipment in the GSM mobile network.

The synchronization elements are SSU in Europe and BITS in North America. These devices lock to the PRC and provide management and distribution of all synchronization signals in the network. The base stations in a GSM network may derive

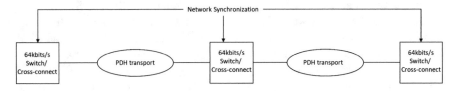

Fig. 9.1 PDH network [4]

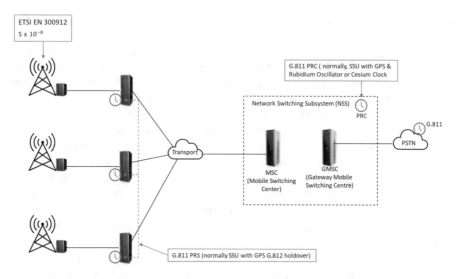

Fig. 9.2 GSM network with clock synchronization

their frequency accuracy from locking a relatively low-performance quartz oscillator integrated in the base station system to a recovered clock signal from the T1/E1 leased line backhaul facility. The figure above shows various standards applicable to such networks for frequency accuracy, for example ITU-T G.811 for PRC and PRS, G.812 for holdover, and ETSI EN 300912 for radio interface.

9.2.2 SONET/SDH Network Synchronization

The need for higher bandwidth and better performance led to the introduction of SONET (Synchronous Optical Network) in North America and SDH (Synchronous Digital Hierarchy) in Europe. However, it took almost 10 years to implement these technologies costing almost $4.5Billion in North America alone. The full implementation of SONET/SDH was continued up until 2000. SONET and SDH network needed end-to-end synchronization. ITU-T G.803 defines architecture of SDH while SONET was defined by ASNI standards ANSI T1.105, ANSI T1.106, and ANSI T1.117. The basic unit of framing for SONET is Synchronous Transport Signal (STS) with STS-1 as the base-level signal at 51.84 Mbps. The STS-1 frame can be carried in an OC-1 signal. The basic unit framing in SDH is Synchronous Transport Module (STM), with STM-1 as the base-level signal at 155.52Mbps. The STM-1 frame can be carried over OC-3 signal. Unlike PDH, SONET and SDH required end-to-end synchronization. If SONET/SDH equipment is not synchronized, then jitter and wander will be generated. Figure 9.3 illustrates SONET/SDH network synchronization needs. The basic principle of SONET/SDH synchronization is that each node must have a highly reliable reference clock, and all nodes

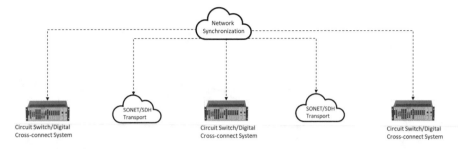

Fig. 9.3 SONET/SDH network requires end-to-end synchronization

should be within a specified level of tolerance to one another. Distributing such a reference clock to each node introduces an intricate problem.

Propagation delays or jitter experienced as a signal passes through a network could introduce phase disparities in a SONET/SDH network. To overcome this SONET/SDH employ a pointer mechanism that allows for any expected phase differences in a network to be accommodated. Pointers provide a simple means of dynamically and flexibly phase-aligning SONET/SDH payload.

The SONET/SDH network is organized into a clock hierarchy levels for synchronization, PRC/SSU and PRS/BITS hierarchy levels are used in SONET and SDH, respectively. Please refer to Chap. 6 for further details.

9.2.2.1   3G Mobile Networks Synchronization

While introduction of SONET/SDH has drastically changed telecom network and improved performance, mobile network system also undergone transformation from derivatives of 2G such as GPRS (2.5G) and EDGE (2.75G) to third generation (3G). Collectively GSM technologies including GPRS and EDGE all rolled under 3G Universal Mobile Telecommunications System (UMTS). International Telecommunication Union (ITU) picked up the standardization of UMTS technologies under the IMT-2000 program. The ITU adopted a limited number of mobile system standards proposed by standardization bodies from all over the world. The UMTS entities are grouped into three domains: Core Network (CN) domain, UMTS Terrestrial Radio Access Network (UTRAN) domain, and User Equipment (UE) domain. The core network is further divided into serving network and home network. The serving network represents the core network functions that are local to the users' access point and thus their location changes when the user moves. On the other hand, home network contains subscription information and databases to which the user is linked by his subscription contract. At radio access level, the UMTS can be divided into two categories UMTS-FDD and UMTS-TDD. For UMTS-FDD, a frequency division duplexing (FDD) technique is used where uplink and downlink transmit simultaneously and use different frequencies, whereas UMTS-TDD that

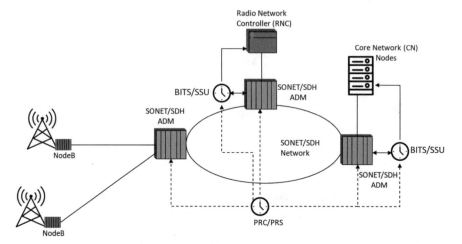

Fig. 9.4 UMTS-FDD mobile network synchronization

uses time division duplex (TDD) technique allows uplink and downlink to use the same frequency but transmit at different times. Both UMTS-FDD and UMTS-TDD relied mainly on SONET/SDH for transport but also PDH and leased lines. For synchronization, all UTMS core network nodes must be synchronized to frequency with an accuracy of 1E-11 or 1×10^{-11} while base stations should be synchronized to frequency with an accuracy of 5E-8 or 5×10^{-8}. Figure 9.4 shows UMTS-FDD mobile network with SONET/SDH backhaul.

Based on whether the backhaul network is SDH or SONET, appropriate clock synchronization is done using PRC/PRS and SSU/BITS, respectively. In UMTS-FDD mode, a single timing reference is not actually necessary within the UTRAN and phase alignment between NodeBs is not essential. However, NodeBs should all be traceable to the same long-term frequency reference (e.g., UTC with the GPS of satellites). In UMTS-TDD mode, common timing reference among NodeBs required to support cell synchronization. As shown in figure below, NodeBs are getting frequency and phase synchronization through embedded GPS receiver which becomes priority 1 feed for synchronization. Frequency feed from SONET/SDH ADM become secondary.

The synchronization among the UTMS-TDD NodeB to one common clock is known as inter NodeB synchronization and essential in UMTS-TDD. Please note, inter NodeB synchronization can also consider frequency input from incoming transmission link as shown in Fig. 9.5.

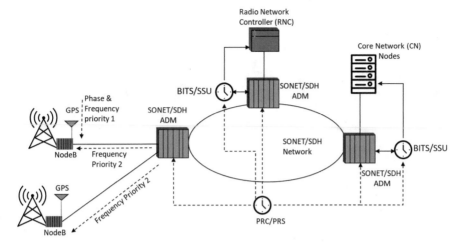

Fig. 9.5 UMTS-TDD synchronization

9.3   TDM to IP/MPLS Network Transformation

Thus far we have discussed how 1G to 3G mobile network transformation happened while digital switching technologies introduced in telecom network allowed mobile network to take leap from 64 Kpbs in 2G to 2 Mbps of bandwidth in 3G at the same time offering data services in the same network. The fourth generation of mobile network was a further leap forward of supporting from 100 Mbps to 1 Gbps bandwidth. Support for such enormous bandwidth was possible due to increased penetration of ethernet technologies in telecom network and transformation of end-to-end telecom networks to IP/MPLS based network from TDM-based network used during 3G era. Although SONET/SDH was considered packet network, recent IP/MPLS network is truly packet-based transport. In some parts of the world, this transformation from TDM to IP based network is still continuing while some service providers create hybrid architecture leveraging benefits of existing infrastructure based on SONET/SDH and extending enhanced services through modern IP/MPLS network. First step towards utilizing both existing TDM and new IP/MPLS based infrastructure is to use Circuit emulation (CEM): it is dubbed as circuit-to-packet migration. Leading vendors such as Cisco and Ciena offer products that allows operators to integrate their existing install base of SONET/SDH, PDH and Ethernet to an MPLS-based network as shown in Fig. 9.6. For example, Ciena's 6500 Packet Transport System (PTS) allows seamless integration of SONET/SDH and other TDM tributaries to MPLS network.

These packet transport systems allow migration of SONET/SDH based telecom transport system and power substation infrastructure. In this example, TDM network for Remote Terminal Unit (RTU) supporting power substation can be easily integrated to IP/MPLS network infrastructure. The packet transport system (PTS) switches support frequency, phase and time synchronization. The PTS switch can

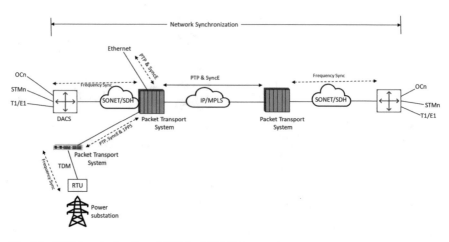

Fig. 9.6 CEM-based migration of TDM to IP/MPLS network

use PTP among themselves as well as SyncE for backup as long as nodes in between support SyncE. A PTS can also take 1PPS for frequency input if needed. Existing SONET/SDH already uses frequency input for synchronization which can innately be used with PTS system. Globally many telecoms started to deploy ethernet for last mile connecting existing 4G cell towers with IP/MPLS infrastructure. I worked on many projects over the years in which both CATV and traditional service providers utilized whitebox ethernet switches to build infrastructures that support mobile transport and FWA (Fixed Wireless Access) for multimedia services. For example, a service provider in South America wanted to utilize its existing 4G infrastructure and extend its service by providing television services (streaming video) to its customer base. To do so, it used existing fiber that was used for SONET ring to create an ethernet ring around its 4G metro cell towers and connected to its EPC (Evolved Packet Core) for transport processing.

Figure 9.7 shows how existing SONET ring was used for ethernet-based MPLS network.

All cell towers were connected to eNodeB that in turn used 10GbE link to 10/100GbE whitebox ethernet switches. All whitebox switches are connected in a MPLS ring and further connected to EPC for mobile network transport processing. The synchronization for mobile traffic was done using both SyncE among whitebox nodes and PTP across the network. The FWA data services implementation did not require precision time accuracy but 4G-TDD did. Since both mobile and FWA transport sharing the same network, it was important to implement proper time synchronization for the common network.

The transformation of SONET/SDH network to IP/MPLS based network allowed further services (e.g., 5G) to be rendered in same network.

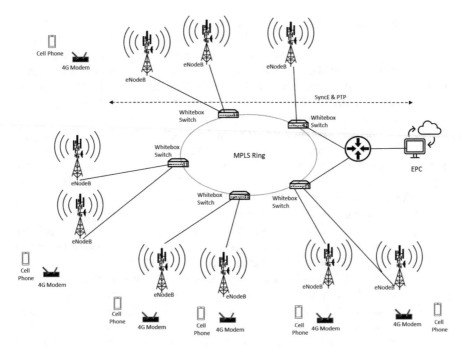

Fig. 9.7   4G Mobile network and FWA design

9.3.1   4G Mobile Network Synchronization

The interest in fourth generation (4G) cellular system has been a topic for sometimes and even before the beginning of transformation of telecom networks to IP/MPLS. The root of 4G network requirements can be linked to formation of third Generation Partnership Project (3GPP) that was established in 1998. For the 3G UMTS, IMT-2000 program set data rate requirements of 2048 kbps for an indoor office, 384 kbps for outdoor to indoor pedestrian environments, 144 kbps for vehicular connections, and 9.6 kbps for satellite connections. However, initial implementation of 3G UMTS did not meet these goals set by IMT-2000, which led to further work and improvement of 3G standards. The combination of High Speed Downlink Packet Access (HSDPA) and the subsequent addition of an Enhanced Dedicated Channel, also known as High Speed Uplink Packet Access (HSUPA), led to the development of the technology referred to as High Speed Packet Access (HSPA) or, more informally, 3.5G [6].

With increase demand for mobile broadband services and quality of service (QoS) needed for high data rates, 3GPP started working on two parallel projects: Long-Term Evolution (LTE) and System Architecture Evolution (SAE). These technologies were intended to design Radio Access Network (RAN) and the network core that included in the 3GPP release 8. The LTE/SAE was also known as Evolved Packet System (EPS) which represented a radical step forward for the wireless industry that aims to provide a highly efficient, low-latency, packet-optimized, and more secure service. At the RAN, both OFDM (Orthogonal Frequency Division Multiplexing) waveforms and MIMO (Multiple-Input Multiple-Output) technologies were introduced while all IP flat network architecture was defined for the transport. The first deployment of 4G LTE happened in December, 2009 at the two Scandinavian capitals Stockholm and Oslo. Later, 3GPP release 10 defined LTE-Advanced that provides among other improvements a higher throughput rate up to 1Gbps in comparison to 300mbps for LTE and the technology is backward compatible to LTE. The LTE can operate both in FDD and TDD mode. We learned from previous discussions that FDD mode of operation in 3G UMTS used frequency input for synchronization, the same is true in LTE-FDD but for LTE-TDD as in UMTS-TDD both phase and time synchronization are required. The backhaul migration of telecom network to ethernet as discussed, has spawned new standardized synchronization techniques, such as synchronous Ethernet (SyncE) from the ITU G.8262 and one from the IEEE called Precision Time Protocol 1588-2008 (PTP).

The SyncE (details of which is presented in Chap. 6) allows carrying PRC input via the Ethernet PHY much the same way T1/E1 carried the traceable clock in the past. This clock provides a very accurate and stable frequency with minimal wander so it easily can be used to synchronize radios in an LTE-FDD base station. SyncE cannot carry a phase component. When it is used by itself, then it's appropriate for FDD-based radio synchronization. For LTE-TDD, PTP and SyncE are normally used for frequency phase and time synchronization. Other technologies such as 1 PPS and 10 MHz are also used for frequency synchronization depending upon network design.

It is to be noted that LTE-TDD and LTE-A must use a time and phase reference input traceable to UTC. As discussed, 2G, 3G, and LTE-FDD need frequency accuracy of 50 ppb at the radio interface. To meet the requirement, 16 ppb frequency accuracy is defined for base station interface to the backhaul network. Table 9.1 shows various frequency and phase requirements for 2G, 3G, and 4G networks according to ITU-T recommendation G.8271.

The work on LTE-Advanced (LTE-A) has introduced some innovative technologies such as carrier aggregation, multiple-input multiple-output (MIMO), coordinated multi-point (CoMP), and enhanced inter-cell interference cancellation (eICIC). The carrier aggregation allows LTE-A to fully utilize the wider bandwidths of up to 100 MHz, while keeping backward compatibility with LTE. It consists of grouping several LTE "component carriers" (CCs) (e.g., of up to 20 MHz), so that the LTE-A devices are able to use a greater amount of bandwidth (e.g., up to 100 MHz), while at the same time allowing LTE devices to continue viewing the spectrum as separate component carriers [6].

Figure 9.8 shows carrier aggregation in contiguous bandwidth (please refer to A) for which an LTE-A user uses up to 100 MHz bandwidth while an LTE user uses 20 MHz in the same band. The carrier aggregation setup in noncontiguous bandwidth shown in the figure (please refer to B) has two LTE users using bandwidths of up to 20 MHz, coexisting with an LTE-A user who is using noncontiguous aggregated bandwidth of up to 100 MHz. It is also possible that, given the proximity of

Table 9.1 Frequency and phase requirements for 2G to 4G networks [7–9]

Mobile network technology	Frequency accuracy for network and air interface	Class level of accuracy (G.8271)	Phase	Note
GSM, UMTS, LTE-FDD	16 ppb/50 ppb	N/A	N/A	
LTE-TDD	16 ppb/50 ppb	3 and 4	1.5 μs for small cell 5 μs for large cell	*End application requirements*: For LTE-TDD wide area base station requirement is 3 μs for small cell (<3 km radius) and 10 μs for large cell (>3 km radius)
LTE MBMS (multimedia broadcast multicast services)	16 ppb/50 ppb	3A	±5 μs	The cell phase synchronization accuracy measured at BS antenna connectors shall be better than 5 μs
LTE-advanced (eICIC, CoMP, and MIMO)	16 ppb/50 ppb	4A, 6A,6B and 6C	±3 μs[a] 260 ns[b] 130 ns[c] 65 ns[d]	Note[a–d]

Note (please refer to ITU-T G.8271):

[a]NR (new radio) intra-band noncontiguous (FR1 only) and inter-band carrier aggregation; with or without MIMO or TX diversity

[b]LTE intra-band noncontiguous carrier aggregation with or without MIMO or TX diversity, and inter-band carrier aggregation with or without MIMO or TX diversity. NR (new radio) intra-band contiguous (FR1 only) and intra-band noncontiguous (FR2 only) carrier aggregation, with or without MIMO or TX diversity

[c]LTE intra-band contiguous carrier aggregation, with or without MIMO or TX diversity. NR (FR2) intra-band contiguous carrier aggregation, with or without MIMO or TX diversity

[d]LTE and NR MIMO or TX diversity transmissions, at each carrier frequency

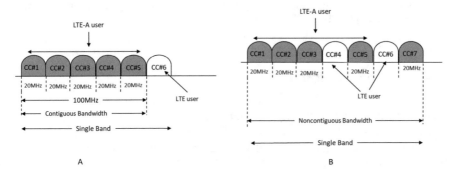

Fig. 9.8 Carrier aggregation with contiguous and noncontiguous bandwidth

LTE-A eNB (eNodeB) and LTE eNB, an LTE-A device has the flexibility of using LTE spectrum and taking advantage of extra bands provided by LTE-A eNB. This requires a setup of coordinated transmission from multiple eNBs enhanced through the MIMO and CoMP schemes available in LTE-A network. While MMO provides capabilities to use multiple antennas at both the transmitter and receiver sides, CoMP allows several geographically distributed nodes to work in coordination while minimizing inter-cell interference to offer improved performance to users within a common cooperation area. The eICIC on the other hand is an interference control technology that prevents inter-cell interference and allows cell-edge UEs (user equipment) in neighbor cells to use different frequency ranges.

While these technologies offer very interesting features, their implementation should be carefully designed. For that, operators must first evaluate their end-to-end network timing and synchronization needs as some of these technologies require phase synchronization engineering rather than classical frequency pulse distribution as evident from Table 9.1.

9.4   Telecom Profile

It is evident from the discussion thus far that future telecom network requires both phase and frequency synchronization which are important and collectively these are called time synchronization. Therefore, to achieve time synchronization in a telecom network both frequency and phase synchronization needed. International Telecommunication Union (ITU) has defined three telecom profiles as follows:

- Frequency Profile: ITU-T G.8265.1
- Phase and time synchronization profile with full timing support from the network: ITU-T G.8275.1
- Phase and time synchronization profile with partial timing support (PTS) from the network: ITU-T G.8275.2

9.4.1   Frequency Profile

ITU-T recommendation G.8265.1 specified frequency profile for use in the PTP network. It was the first telecom profile written with consideration of 3G networks and its migration to packet network. The G.8265.1 is based on G.8265 that defines frequency delivery in a packet network. The G.8265.1 frequency profile aims to allow interoperability between PTP master clock and PTP slave clocks for frequency distribution. This means both master and slave must interoperate with each other in a vendor agnostic way. The profile also defines message rates and parameters for frequency distribution over Ethernet, IP, and MPLS networks while supporting interoperability with existing SONET/SDH infrastructure. Another important feature of this profile is that a slave and master supporting this feature allows for slave to switch from one grandmaster or packet master clock to the alternate grandmaster. To operate in this profile, a PTP domain needs to be established using unicast messages with only one grandmaster or packet master clock per PTP domain and domain number should be same for all clock within that domain. The G.8265.1 profile also supports one-step and two-step PTP mode which was discussed in Chap. 8 and PTP transport over UDP datagram based on IPV4. The message rate should be as follows:

1. *Sync messages (including Follow_Up)*: 128 packets per second with minimum rate of one packets every 16 s.
2. *Delay_Req/Delay_Resp messages*: 128 packets per second with minimum rate of one packets every 16 s.
3. *Announce messages*: eight packets per second with minimum rate of one packet every 16 s. The default value is one packet every 2 s.
4. Signaling messages and Management Messages: no rate specified in G.8265.1 and under further study.

This profile does not use default BMCA (please refer to Chap. 8 for details) instead uses a master selection process (please refer to ABMCA process presented in Chap. 8 for further details) based on the following parameters from a locally provisioned master list according to the priority assigned:

• quality level,
• packet timing signal fail (PTSF-lossSync, PTSF-lossAnnounce, PTSF-unusable),
• Priority.

The quality level is carried in the clockClass attribute by the Announce messages of the candidate master as discussed in Chap. 8. The following table shows mapping of quality level to ClockClass Attributes (Table 9.2).

To further understand how this table works, please refer to Chap. 6 section "BITS and SSU." This table is also presented in that chapter with detailed discussion of SSM quality level to PTP mapping.

9.4.2 Phase and Time Synchronization Profile with Full Timing Support

The full timing support profile for phase and time synchronization in a PTP Network defined by G.8275.1 uses boundary clocks to extend time support across the network and hence, the performance is not affected by the PDV of the network. The profile also supports ordinary clock (OC) and transparent clock (TC) as defined in IEEE1588. Additionally, the use of synchronous ethernet is mandatory with this profile. Figure 9.9 depicts typical network setup with G.8275.1.

As shown in the figure above, boundary clock is used in the DCSG or Cell Site Router (CSR) to extend the reach of timing support to the fronthaul endpoints. Multicast messages are used for all messages using this profile as per the PTP mappings defined in IEEE 1588-2008 Annex F. Both the non-forwardable and forwardable multicast addresses are permitted (01-80-C2-00-00-0E and 01-1B-19-00-00-00,

Table 9.2 SSM QL to PTP ClockClass Attribute mapping [9]

SSM QL	ITU-T G.781			PTP ClockClass
	Option I	Option II	Option III	
0001		QL-PRS		80
0000		QL-STU	QL-UNK	82
0010	QL-PRC			84
0111		QL-ST2		86
0011				88
0100	QL-SSU-A	QL-TNC		90
0101				92
0110				94
1000	QL-SSU-B			96
1001				98
1101		QL-ST3E		100
1010		QL-ST3/ QL-EEC2		102
1011	QL-SEC/ QL-EEC1	QL-SEC		104
1100		QL-SMC		106
1110		QL-PROV		108
1111	QL-DNU	QL-DUS		110

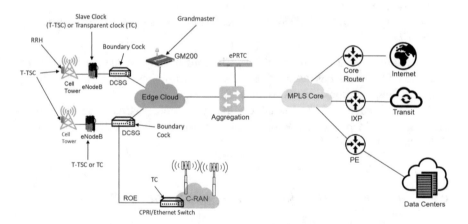

Fig. 9.9 Typical 4G/LTE-A Telecom network with full timing support using G.8275.1

respectively). The multicast address is set by configuration on a per-PTP port basis, but both non-forwardable and forwardable multicast addresses must be accepted by the PTP port. The PTP domain configuration supports a default value of 24 with a range 24 to 43 supported for configuration.

The messages types and rates used in this profile are:

- Sync messages (including Follow_Up): nominal rate of 16 packets per second.
- Delay_Req/Delay_Resp messages: nominal rate of 16 packets per second.
- Announce messages: nominal rate of 8 packets per second.

The profile uses ABMCA as specified in IEEE1588 (please refer to Chap. 8 for details) which allow manual provisioning.

9.4.3 Phase and Time Synchronization with PTS Profile

This profile supports two deployment cases: first one is for the Assisted Partial Time Support (APTS) and the second is PTS in which a PTSC (Partial Timing Support Clock) recovers time/phase using PTP. In the first case, GNSS PRTC support would be collocated with APTSC (Assisted Partial Time Support Clock) and PTP will be used as backup. The strong on-path support for G.8275.1 FTS (full timing support) profile limits its applicability to many greenfield deployments or in cases where network is simple and modern enough to allow for a deep re-engineering. This limitation of FTS led ITU to add this new profile for telecom environment. To understand as to why G.8275.2 profile is relevant, let us explore the advantage and disadvantages of PTS and APTS. The PTS allows a more relax set of requirements to the network than FTS. The important point to note is that the former does not require all transit nodes from the grandmaster to the slave to be PTP aware. In other words, FTS becomes PTS if at least one T-BC is replaced by a non-PTP aware device. Let's take the example of network presented in Fig. 9.9 for FTS and apply the concept of PTS as shown in Fig. 9.10.

Please note DCSG is replaced by standard whitebox that is PTP unaware. In this case, depending upon network setup eNodeB can be a transparent clock or a slave clock. Assuming that a Radio Remote Head (RRH) or Radio Unit (RU) behind eNodeB is using T-TSC, it would be better to use eNodeB as transparent clock. Nonetheless, there is no boundary clock here in between eNodeB and the grandmaster. This type of deployment may reduce TCO since the CSR can be replaced by low-cost whitebox. However, it is expected that these whitebox or PTP-unaware

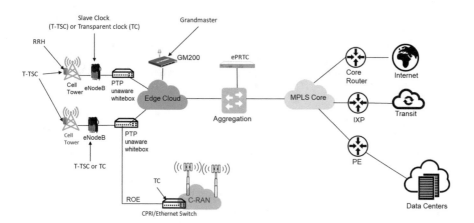

Fig. 9.10 An example of PTS network

devices are able to prioritize IP/UDP traffic for PTP. It is to be noted that the total time error budget from PRTC in grandmaster to endpoint is 110 ns or 1.1 μs.

The APTS architecture evolves from deployments that rely entirely on GNSS. The advantage of these architectures is that they do not require any synchronization support from the network but on the other hand they require massive GNSS facility installation at the network edges. Such deployments are vulnerable to GNSS signal jamming or spoofing but some vendors such as Trimble are already offering solutions to overcome GNSS jamming and spoofing issues through dual band timing device that provides anti-jamming and anti-spoofing capabilities. Let us further explore the PTS deployment discussed earlier and apply APTS concept to it as depicted in Fig. 9.11.

In this scenario, RRH and eNodeBs are not using PTP and they are receiving UTC through RES720 timing module which can be integrated within those devices. The eNodeB in this scenario may use PTP as backup in case of GNSS failure. The RES720 timing module presented in this diagram provides dual band capabilities for GNSS meaning it can fall back from L2 to L5 signal if one signal fails or the signal is jammed. The timing module also provides anti-jamming and anti-spoofing capabilities as well. However, it is recommended that maximum time error for the network is 1350 ns or 1.35 μs with or without GNSS. The APTS has emerged as a GNSS-assisted architecture that uses PTP for backup rather than physical layer synchronization. The main advantage is that full timing support from the network is not required.

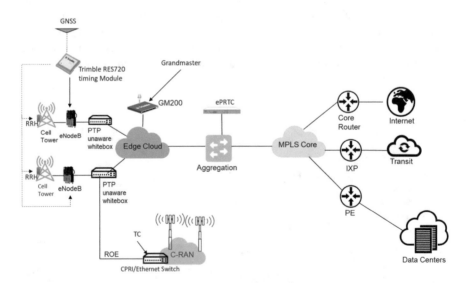

Fig. 9.11 APTS Network using Trimble's RES720 timing module

9.5   The Telecom Network Trends

It is obvious that telecom network from fronthaul to the core is transforming rendering the transport systems to more software defined and intelligent. Along the way the architectural changes that are induced by 5G technologies transformed it to a versatile and service-oriented infrastructure. Never before it is perceived so encompassing and the transformations that occur in every industrial vertical is just tip of the iceberg. For it, telecom cloud is an enabler of many possibilities that stemmed from pervasive service-oriented on demand connectivity. Central to this transformation is the notion of open networking, I say this, because the notion of collective innovation has changed the networking world which serves the backbone of telecom infrastructure. So, the challenges of connectivity have somewhat been addressed, what remains is the collective innovation of network decomposition as it applies to Telecom cloud. From this observation, we can state that open networking is an enabler of transformation in the connectivity infrastructure. But for it to make the headway in the transformation towards the future of telecom cloud, other catalysts were needed and those include works related fifth generation (5G), softwarization, network decomposition, and virtualization. Figure 9.12 attempts to summarize various forces of technologies at work that collectively defines ubiquitous connectivity for which telecom infrastructure is central.

During 2015, I first came to experience the transformation that can be induced by ubiquitous connectivity and telecom infrastructure. As part of the scores of RFIs (Request for Information) from operators, interconnect exchanges and hyperscale data centers around the world in which we participated an obvious objective was the reduction of CAPEX/OPEX but more importantly to create dynamic service-oriented infrastructure. Central to their interest was understanding the maturity of technologies that stemmed from open networking and often called whitebox. This historical perspective is important, since the said open networking has gradually

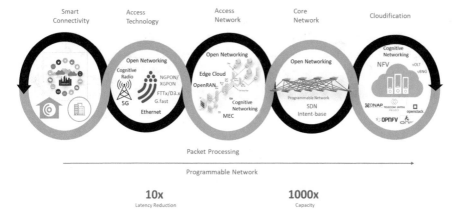

Fig. 9.12 Trend in telecom network technologies

expediated many technological advents that today we take for granted, e.g., SDN (Software Defined Networking). What started as an experiment by hyperscale data center to transform their network unfolded as collective innovation in the form of Open Networking. Hyperscale data center saw a phenomenal growth of bandwidth and applications thanks to the internet, social networking, and ubiquitous connectivity that demanded more insights and enhanced experience enabled by data services. In parallel, system virtualization and advances in computing created great potential for a plethora of data services, data harnessing and applications. The problem was black box issues and cost of both server and networking systems, but server vendors were more forth coming in creating open system decoupling hardware and software allowing collective innovation for each and driving the cost down in process. However, network systems remained vendor locked and expensive, that led hyperscale data centers to look for alternative and one such alternative was to specify their own hardware and procuring these from ODM (original design manufacturer) suppliers. By 2016, hyperscale data centers started procuring hardware from ODM suppliers and created NOS (Network Operating System) for those networking hardware. The hardware, however, came with firmware and thanks to the support from open-source community, a standard API for NOS implementation. These barebone hardware (actually these were not so barebone as stated) were known as "bare metal" switches. By late 2016, a number of NOS were available for bare metal switches, some from open source while others from startups who wanted to become NOS vendor. These industry undertakings created whitebox (bare metal switch with NOS) and paved the way for software defined networking (SDN) to take root as various APIs and protocols enabled the separation of control plane and simplified automation, management, and orchestration of network elements.

Figure 9.13 depicts how open networking enabled the future of networking including telecom infrastructure. The notion of open networking gives birth to disaggregation or decoupling concept that separates hardware and software while ensuring seamless integration. This concept of disaggregation and 5G induced network decomposition allowed collective industry innovation of OpenRAN that

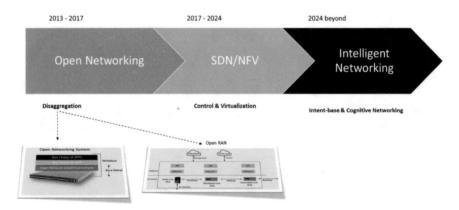

Fig. 9.13 Network and Telecom infrastructure transformation led by Open Networking

innately created possibilities for functional decomposition of x-haul (fronthaul, midhaul, and backhaul of mobile network infrastructure). The experimentation with decoupled system enabled by disaggregation paved the way for further research on softwarization of network and software defined networking (SDN). With decoupling of control plane and notion of decomposition, industry came together to innovate on the portability of network elements what is known as VNF (Virtual Network Function). These fundamentals are changing telecom infrastructure as we know it and rendering to what I stated before, "the telecom cloud."

The possibilities can be further extended to intelligent networking for which we are already observing deployment of networks that are policy or intent based. The concept is known as intent based networking (IBN). The IBN utilizes artificial intelligent (AI) and machine learning (ML) to prescribe and perform routine tasks, set policies, respond to system events, and verify that goals and actions have been achieved.

However, IBN by itself cannot create the "network state" for self-healed and self-organized network. A more dynamism and innate intelligent needed to be part of the network which perhaps can be induced by AI/ML and implemented in the "state machines" of every network element. This would be the prelude to what is known as cognitive networking, a form of intelligent networking and perhaps will redefine the future of telecom network nay the telecom cloud.

9.6   The Era of 5G

As the network is undergoing the transformation, so do the mobile communications network. Thus far the transformation from 1G to 4G has been leapfrogging but that is about to change with fifth generation (5G) mobile communication technologies. Defined by 3GPP releases 15 and 16, 5G innately divides telecom infrastructure into two parts: RAN and packet core. To cater the demand of future, we need a reliable and more stronger mobile connectivity that facilitate pervasive connectivity and whopping bandwidth for service on demand. That's where 5G comes in. It offers wireless speeds comparable to today's wired broadband, while delivering better energy efficiency than modern 4G networks.

As the next step in the evolution of mobile communication, 5G aims to enable ubiquitous connectivity, in other words provide connectivity everywhere for any kind of device that may benefit from being connected. 5G will support a wide range of new applications and use cases, including smart homes, traffic safety, critical infrastructure, industry processes, and very-high-speed media delivery. And it will accelerate the development of the Internet of Things.

This will be one of the most significant technological transformations of the twenty-first century, with implications for nearly all sectors of society and industries. Ultimately, 5G with collective of technologies such as Artificial Intelligence and automation have great potential to improve the way we live and work.

9.6.1   New Network Model

As an enabler 5G has the potential to break the barrier of monotony in connectivity and network infrastructure. It fits well with software-centric network disaggregation approach that innately removes the constraints to build dynamic and application-driven environment. Interestingly, it brings possibilities of cloudification and automation to help operators meet these new application and operational demands. Operators will reap the benefits of true cloudified networks that are harmonized with a common feature set across all target markets. With the onset of a new software-defined architecture, the supply chain for mobile network infrastructure deployment changes at a fundamental level supporting unprecedented level of versatility, allowing operators to combine best-in-class functions to offer enhanced service. It will also help operators evolve services as needed to address market needs.

Since the new architecture embraces software-centric network disaggregation approach, they promote automation and service versatility. A prime example would be 5G use cases that target enterprise, industrial automation, autonomous vehicle, smart city, smart healthcare, smart grid, and numerous other industry verticals. These services should be supported using softwarization, network slicing and virtualization, automation and orchestration of IBN for the service intents. Additionally, many of the software-defined functions will occur in virtualized environments at or near the edge of networks, which enables support for a newer breed of low-latency services defined in edge computing or multi-access edge computing (MEC) [10].

9.6.2   5G Mobile Network Architecture

As discussed earlier, 5G break down monolithic mobile infrastructure into functional decomposition of control and user planes allowing more flexibility beyond hardware and software disaggregation. The first step towards this decomposition is to split radio signal processing stack. Figure 9.14 depicts radio signal processing stack split options specified for 5G. The gNB (gNodeB) function that processes entire radio protocol stack is split into eight different options for deployment: Radio Unit (RU) comprising RF and LPHY (lower) PHY functions, Distributed Unit (DU) comprising Upper PHY (UPHY) to ULRC (Upper Radio Link Control) and Centralized Unit (CU) comprising PDCP (Packet Data Convergence Protocol) and RRC (Radio Resource Control).

Some operators are also trying to implement similar concept to augment their existing 4G install base by splitting eNodeB (eNB) to RRH (Radio Remote Head) and Base Band Unit (BBU) as shown in the figure above. The concept of virtual BBU (vBBU) that allows virtualization of upper layer radio protocol stack functions to a server is becoming a common type of deployment for future 4G network decomposition. This concept of network function virtual (NFV) creates flexibility and scale up capabilities for operators. Thus, the value of micro data center at the

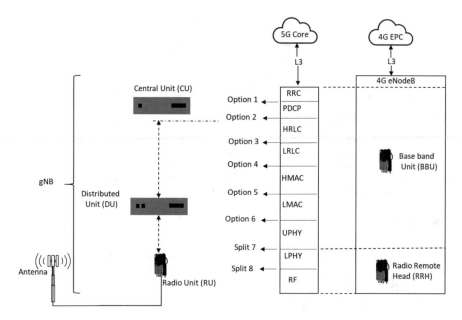

Fig. 9.14 Radio signal processing stack and suggest 5G radio stack Split

edge or what known as "edge cloud" becomes extremely useful for operators. Once this network decomposition is complete, it will allow them to bring contents and other services to the edge. However, this network decomposition also allows other service providers (those are not traditional operators) to offer FWA (Fixed Wireless Access) and mobile network services to end users. Suddenly, what we knew as monolithic infrastructure is fluidified into VNF (Virtual Network Function) commodities that can be deployed at will and scale up as needed to offer service on demand.

Welcome to the world of telecom cloud: possibilities are endless.

9.6.3   OpenRAN

The discussion about 5G remains incomplete if we do not highlight the development in OpenRAN initiatives of O-RAN Alliance and Telecom Infrastructure project. In fact, OpenRAN is at the forefront of rendering 5G services to more accessible. The open networking or more specifically the disaggregation concept we discussed earlier that shaped the networking industry has found its way to Telecom fronthaul. Thanks to the initiative of many industry forums including O-RAN Alliance, Telecom Infrastructure Project (TIP), Open RAN Policy Coalition and Small Cell Forum, Network decomposition of 5G is becoming reality through collective innovations that give birth to OpenRAN.

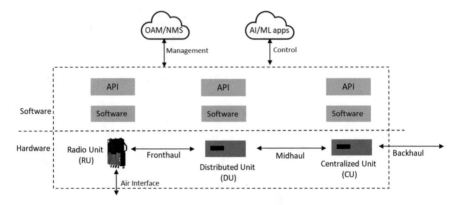

Fig. 9.15 OpenRAN architecture

The OpenRAN allows decoupling of hardware and software and specifies standard APIs for control, management, and orchestration. Figure 9.15 depicts conceptual model of OpenRAN as defined by TIP.

The key tenets of OpenRAN are as follows:

1. Disaggregation of RAN hardware and software allowing vendor agnostic deployment on GPP (General Purpose Processor) based platform.
2. Standard open interface between components (RU, DU, CU, etc.).
3. Architectural flexibility using radio protocol stack split options.
4. Multi-vendor solutions enabling a diverse ecosystem.
5. Innovation and adoption of new technologies such as containerization, integration of AI/ML, etc.
6. Supply chain diversity.

While OpenRAN create flexibility through disaggregation of RAN components, it also allows tighter control of standard interface of communication between the component to foster vendor agnostic solutions through open interface. Additionally, OpenRAN fosters collective industry innovation for telecom fronthaul infrastructure.

9.6.4 5G Sync Plane

The 5G radio access networks (RANs) are getting a major upgrade which can extend coverage and boost performance for end users services. With continued allocation of spectrums by FCC, the deployment of 5G New Radio (NR) can now benefit from availability of wide range of spectrums in low-band FDD spectrum, in combination with mid- and/or high-band TDD spectrums. This will be a game changer for 5G network operators. Figure 9.16 depicts 5G spectrums in both low-band FDD and potentially mid- to high-band TDD.

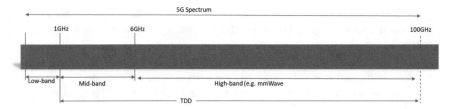

Fig. 9.16 5G low-band FDD and mid- to high-band TDD spectrums

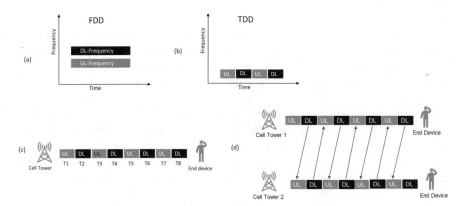

Fig. 9.17 Difference between FDD and TDD transmission

With combination of low-band and high-band spectrum, 5G can overlay mmWave RAN solutions on top of low-band spectrum in a dynamic spectrum sharing while overcoming coverage and performance. However, with TDD spectrum sync plane issues become a major issue and require careful design of 5G RAN to overcome sync plane challenges.

9.6.4.1   Time Division Duplex (TDD)

Fundamentally, TDD differs from FDD in transport; the former needs timeslot allocation to share same frequency for downlink and uplink where the latter uses different frequency for uplink and downlink. As shown in Fig. 9.17, FDD transmission is represented by (a) and TDD is represented by (b). It is obvious from the diagram that TDD-based transport requires stringent time synchronization since uplink and downlink share same spectrum but different timeslots for each. Therefore, synchronization in radio transmission path as represented by (c) and for that matter entire fronthaul to backhaul requires adequate time synchronization. Failure to do so will result in network performance issues. This may get further complicated when the same operators use multiple cell towers for TDD transmissions to the end devices. As represented by (d), Uplink (UL) and Downlink (DL) transmission may suffer from channel interference if entire fronthaul including the radio endpoints within a

CoMP solution are not properly synchronized. This scenario gets further complicated when multiple operators share the same frequency band operating in the
same area.

In this case, similar to the situation described as (d), downlink will interfere with
uplink transmission causing significant capacity issues. Henceforth, synchronization of frequency and phase is very important in TDD and for 5G it is must.

9.6.4.2   Time Error (TE) Budget

According to ITU-T recommendation G.8271.1, a mobile network must adhere to a
time error budget of 1.5 µs for the network. The error budget should be considered
from end application to PRTC at T-GM. It is to be noted that end applications may
introduce ±150 nanoseconds to the network. The link asymmetry compensation
is 250 ns.

Between the Radio unit or RRH that are using same DU or BBU must have ≤260
nanoseconds of time error budget. The grandmaster that acts as PRTC should have
maximum absolute TE < 100 ns at PRTC output interface and reference point at
base station the max absolute TE should be <100 ns (please refer to a point between
node 5 and node 6a, b, c in the figure below). This should be the point where a CU
terminates fronthaul connection.

If we consider the entire chain of T-BC from node 1 to node 5, the max absolute
TE should be <1.1 µs. Figure 9.18 depicts relative time error budget for fronthaul
network design.

Additionally, grandmaster to base station there should not be more than 5 boundary clock (T-BC). Some of the tier 1 deployments only 2 T-BCs are used as shown
in Fig. 9.19.

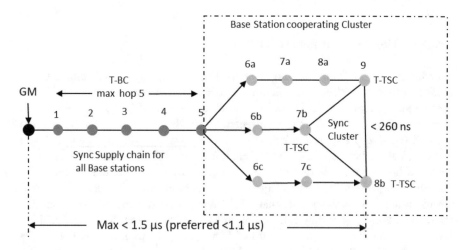

Fig. 9.18   ITU-T G.8271.1 a max time error budget of 1.5 µs for mobile network

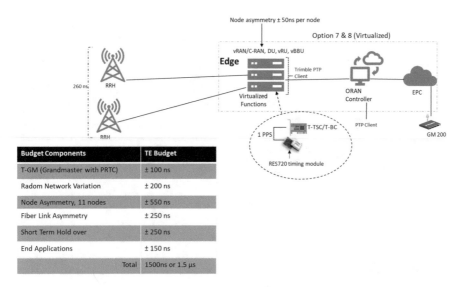

Fig. 9.19 Typical 5G deployment with max 2 T-BC node

The DCSG (Disaggregated Cell Site Gateway) that sits between CU and DU extends and allows many DUs to be connected to CU. The DCSG implements T-BC through embedded T-BC module, herein Trimble GM310. The RUs may integrate timing module such as RES720 for T-TSC (slave clock). The time error budget estimate varies depending upon network design, however estimating per node asymmetry, link asymmetry and T-GM/T-BC level TE allows you to adequately calculate the anticipated TE for the network design under consideration. It is suggested such calculation is done aforehand to eliminate synchronization issues.

Figure 9.20 depicts a typical virtualized fronthaul deployment in which vBBU or DU are stacked together to serve many RRHs or RUs depending upon network setup from an edge cloud. In such scenarios, system integrator may like to integrate timing module with PTP NICs to provide frequency input in-line with APTS profile configuration which we discussed before. This will provide a sense of resilient timing for the deployment. Nonetheless, when it comes to TE budget for the network please refer to the TE budget estimate provided in the figure above.

Last but not the least, please consider resilient timing solutions for your network. It is a mandate now that all critical infrastructure providers must provide resilient timing to their network in case of GNSS failure. It would thus be a suggestion to consider integrating timing module fronthaul devices such as RU, DU, and CUs for resilient UTC input. The dual frequency timing module such as RES720 also supports anti-jamming and anti-spoofing capabilities for the GNSS PRTC input. Additionally, IEE1588-2019 security features are also essential to protect against attacks on PTP.

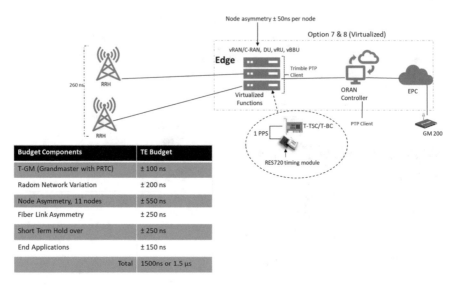

Budget Components	TE Budget
T-GM (Grandmaster with PRTC)	± 100 ns
Radom Network Variation	± 200 ns
Node Asymmetry, 11 nodes	± 550 ns
Fiber Link Asymmetry	± 250 ns
Short Term Hold over	± 250 ns
End Applications	± 150 ns
Total	1500ns or 1.5 µs

Fig. 9.20 Total time error budget

9.6.4.3 CBRS Synchronization

The Citizens Broadband Radio Service (CBRS) is a new 5G technology and nothing to do with Citizens Band (CB) radio that truckers use for two-way communications using 40 select channels within 27 MHz in the USA. The CBRS can be considered "enterprise 5G" similar to private LTE that allows an entity to create their own private 4G networks. Unlike its predecessor which uses licensed 900 MHz, spectrum CBRS (better known as enterprise 5G or private 5G) uses unlicensed 3.5 GHz TDD shared spectrum. A CBRS network can simultaneously host applications that have varied requirements: URLL (Ultra Reliable Low Latency) applications such as robotics or vehicle remote control, which require the highest priority and reliability can be hosted with video monitoring or video communications applications that require more bandwidth but are less sensitive to latency. Additionally, voice services can also be offered for both stationary and mobile users. Enterprises in many different industry verticals such as healthcare, corporate campuses, hospitality, military, manufacturing, transportation, utilities, and supply chain can benefit from CBRS technologies. The CBRS allows both license and unlicensed use of 3.5 GHZ TDD spectrum. Figure 9.21 shows CBRS spectrum and its tiered users. CBRS spectrum is a 150 MHz band that is shared by three distinct tiered users.

The tier 1 users (3550–3700 MHz) include federal users including military and coastal locations, etc. The tier 2 users include fixed satellite services (FSS) and wireless providers with grandfather wireless broadband services license. The FSS uses 3600–3650 MHz while wireless providers with grandfather license use 3650–3700 MHz. The tier 1 and tier 2 users are known as incumbent users in CBRS spectrum allocation. The tier 2 users are granted the license under Priority Access

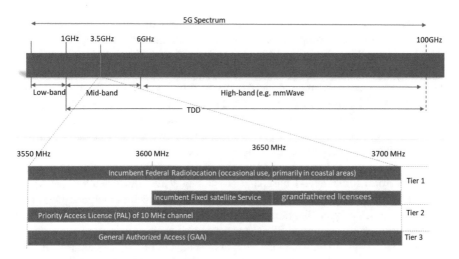

Fig. 9.21 CBRS Spectrum and tiered users

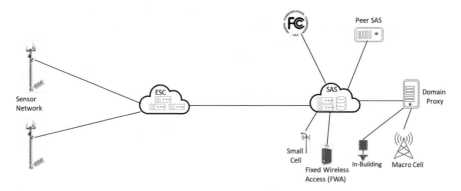

Fig. 9.22 The SAS operation and connectivity with various systems

Licenses (PALs) which is issued in county-by-county basis and up to seven licenses will be granted per county. Each PAL consists of a 10 megahertz channel within the 3550–3650 MHz band with 10-year renewable licenses term.

The tier 3 users granted GAA (General Authorized Access) licensed-by-rule to permit open, flexible access to the band for the widest possible group of potential users. The GAA user may operate in the entire CBRS spectrum but must not cause harmful interference to incumbent access users or priority access licensees and must accept interference from these users. The way to ensure no interference is to implement SAS (Spectrum Access System). The SAS is an automated frequency coordinator that manages spectrum sharing on a dynamic, as-needed basis across all three tiers of access. Figure 9.22 shows how SAS works for different deployment.

The SAS dynamically manage that spectrum use by accessing an Environmental Sensing Capability (ESC) network, Peer SAS and FCC in order to comply with

FCC regulations governing access priorities in areas of military operation. The CBRS is implemented using CBSD (Citizen Broadband Radio Service Device) and SAS but it can also extend to RAN Controller and EPC depending upon deployments. The following diagram shows a mix of Private LTE and CBRS deployment. Many vendors are offering products that can provide a mixed Private LTE and CBRS services for enterprises. As depicted in Fig. 9.23, top of the building LTE-A/CBRS Access Point (AP) provides both license and unlicensed band spectrums allowing voice and broadband services simultaneously. This service can be further extended throughout the building using "Radio Points."

The "Radio Points" transmit and receive radio frequency (RF) signals over the air and perform some layer-1 baseband processing. These devices allow cell virtualization, joint transmit/receive and user location awareness for emergency services. The setup depicted in the diagram below was deployed in an enterprise building and a single grandmaster provided PRTC services to all radio and access points for the campus. Such deployments are also seen in hospitals to provide essential LTE coverage along with reliable broadband services. The devices in such setup are managed by vendor provided tool such as device management system (DMS).

Where CBRS frequency access is required, controller connects to SAS server to allocate appropriate frequency clearance. It is done by verifying spectrum usage and compliance requirements from the SAS server.

For time synchronization, apart from TDD synchronization requirements all CBSDs are subject to same synchronization requirement as LTE eNodeBs. Therefore, time error budget must calculate per device asymmetry, link asymmetry, and overall end-to-end network time budget of 1.5 µs. Frequency synchronization interfaces (SyncE and 1PPS, etc.) for T-BC and CBSDs are not important and

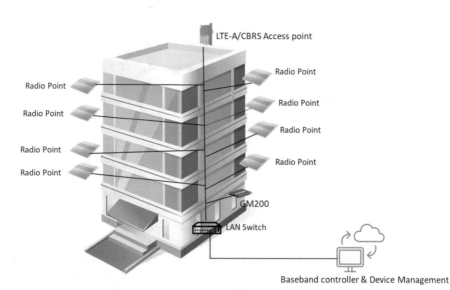

Fig. 9.23 Private LTE and CBRS mixed deployment

should be a good to have features. It is to be noted that all PTP-unaware nodes must prioritize PTP traffic with highest priority to maintain TE budget for the network. Careful consideration is needed for jitter and PDVs.

9.7   Telecom Cloud and Synchronization

From the discussions in this chapter, it is clear traditional telecom network setup is no longer serving the need of modern era. Transformation of mobile network is on the offing and many operators are exploring ways to virtualize network to provide bandwidth on demand and enhanced experience to end users including enterprises. Telecom cloud allows intent-based, cognitive and software-defined network and compute resources to be allocated on demand enabling cloud services for mobile communications, broadband, URLL services, and data/content processing. From technical perspective, it requires data center resources to deploy and manage a mobile phone network with data transfer capabilities by carrier companies in pro-duction operations at scale.

Several customers that I worked with have taken steps to cloudify their networks and utilized micro data centers like setup for distributed edges. Figure 9.24 depicts one of the largest operators who implemented a virtualized mobile network that can be considered as the prelude to Telcom Cloud.

Synchronization for this setup included hundreds of T-BCs for the Far Edge where thousands of DUs are deployed. Despite the careful design, sync plane issues were one of the major concerns for the network due to asymmetry and thus required asymmetry corrections. Therefore, PTP slaves and boundary clocks must provide asymmetry corrections in addition to security features that are de facto in such envi-ronment. Given that telecom cloud most likely to implement the decentralized data center resources for processing at "Far Edge" and "Edge Aggregation," it may ben-efit from Time Synchronization as a Service (TSaaS). As the name implies, TSaaS

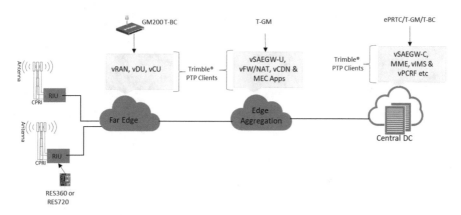

Fig. 9.24 Typical Telcom Cloud setup

extends traceable UTC to virtual endpoints through cloudification of PRTC services. This will be particularly useful for telecom cloud for NFVs environment including various MEC (Multi-access Edge Computing) applications.

References

1. Alsharif, H. M., & Nordin, R. (2017). Evolution towards fifth generation (5G) wireless networks: Current trends and challenges in the deployment of millimetre wave, massive MIMO, and small cells. *Telecommunication Systems, 64*, 617–637. https://doi.org/10.1007/s11235-016-0195-x.
2. Gardner, F. M. (2005). *Phaselock techniques* (3rd ed.). Hoboken, NJ: Wiley.
3. Lin, J. (2018). *Synchronization requirements for 5G. An overview of standards and specifications for cellular networks*. IEEE Vehicular Technology Magazine.
4. ETSI EG 201 793. (2000). *Transmission and Multiplexing (TM); Synchronization network engineering*. ETSI EG 201 793 V1.1.1 (2000–10). European Telecommunications Standards Institute.
5. ETSI EN 300 912. (2000). *Digital cellular telecommunications system (Phase 2+); Radio subsystem synchronization* (GSM 05.10 version 7.1.1 Release 1998). European Telecommunications Standards Institute.
6. Akyildiz, F. I., Gutierrez-Estevez, M. D., & Reyes, C. E. (2010). The evolution to 4G cellular systems: LTE-advanced. *Physical Communication, 3*(4), 217–244.
7. Microsemi. (2014). *Timing and synchronization for LTE-TDD and LTE-advanced mobile networks*. Whitepaper. Microsemi Corporation.
8. Nuss. (2014). *TD-LTE and LTE-advanced networks need correct time and phase data*. ElectronicDesign, Endeavor Business Media, LLC.
9. G.8271. (2020). *Time and phase synchronization aspects of telecommunication networks*. ITU-T G.8271/Y.1366 (03/2020). International Telecommunication Union.
10. Cisco. (2020). *Reimagining the end-to-end mobile network in the 5G era: Rakuten finds success through disruptive thinking and actions*. Cisco Systems, Inc.

Chapter 10
Synchronization for Smart Grid Infrastructure

10.1 Introduction

Over a century, power system went through tremendous changes to provide reliable and steady source of energy. The power system and for that matter power grid is considered a marvel of reliability yet reliability principle puts a limit to continuous 100% uptime. There are various sources that could cause abnormalities in power grid operations. These abnormalities may stem from internal and external causal factors. To keep the power grid reliable and operation 100% of the time, monitoring of power quantities in real time is essential and this requires both integration of new technologies and robust implementation of ICT (Information Communication Technologies) and grid-wide security mechanism to protect it against attacks. Smart grid initiatives take all these context into account while suggesting a time-synchronized architecture for interconnectivity, operations, and control of power grid. Without time synchronization, it would be difficult to visualize, analyze, and take preventive actions of any anomalies of power Grid. In a workshop arranged by NIST (National Institute of Standard and Technologies), grid operator identifies most challenges yet essential requirements for real-time monitoring of power quantities, these include TWFL, use of synchrophasor technologies, WAMS, event timing, and reconstruction to operation and control of digital substations. All these functions require precision time synchronization. This chapter discusses synchronization needs from historical perspective and explores its importance in the operation of modern "next generation" smart grid. Additionally, the chapter specifies end-to-end time synchronization requirements for power system while discussing various power profiles that needed to be implemented for optimal operations of power systems.

D. D. Chowdhury, *NextGen Network Synchronization*, https://doi.org/10.1007/978-3-030-71179-5_10

10.2   History and Evolution of Power Grid

The power grid is an important critical infrastructure which has raised quality of life of all people around the world. Not a day can gone by without electricity. The journey of present-day power grid began with isolated power generation systems across the world starting in the 1870s. The growth and unification of the systems into an interconnected AC (alternating current) can be traced back in history of the development of first alternator by Hippolyte Pixii in 1835. Electricity originated as Direct Current (DC) and scientists at the time were more interested in DC (direct current), but once the advantages of AC power were realized, it became the world standard for electricity. The AC power changes its polarity between positive and negative easily with a transformer, making it more efficient for transmitting power over long distances. This makes AC power more economical for distribution of electricity to the homes around the world.

In 1880s, industrialists and inventors wanted to find ways to power streetlights and homes and that led to "war of currents" between AC and DC power. In the early 1880s, Thomas Edison started first large-scale power utility services in lower Manhattan eventually serving 6 square miles. This distribution of power was DC based and beginning of power distribution system in the USA. First methods used to power both DC and AC generation plants were coal-fired steam engines and hydroelectric power [1]. Figure 10.1 depicts the timeline of distributions and commoditization of power in the USA.

The power grid in the USA evolved from small-rated centralized power stations, such as the historical Pearl Street Station in Manhattan, NY (such power stations supplied DC electricity to relatively smaller-rated loads in the late nineteenth century) to the interconnected AC system of present day that crisscrosses North America [2]. The last two decades saw a steady growth of distributed generation, with plans for higher penetration of renewable energy sources, and the policies on electricity distribution have been supporting needs for the "smart grid" for many reasons that include increased reliability and resiliency, faster restoration after disruptions, more information and energy management for consumers, easier integration of renewable energy, and enhanced security and protection.

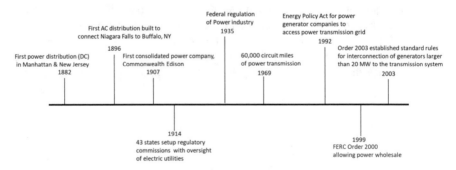

Fig. 10.1 The distribution and commoditization of power in the USA

The conceptualization of techniques to improve the intelligent interaction of distributed power assets for smart grids emerged in the 1980s as a call to modernize the grid, allowing deeper penetration of alternative and renewable energy sources. The first references to the term "Smart Grid" were provided around 2004 by Amin [3], Amin and Wollenberg [4]. Many definitions of smart grid exist, and they collectively define the characteristics or capabilities it enables. A better definition provided by Electric Power Research Institute (EPRI) is as follows:

> A smart grid is one that incorporates information and communications technology into every aspect of electricity generation, delivery, and consumption in order to minimize environmental impact, enhance markets, improve reliability and service, and reduce costs and improve efficiency [5].

The smart grid applies digital technologies to the sensing, communication, computing, control, and information management functions of the power grid. This integration of digital technologies to power grid's electromechanical system evolved over decades beginning the introduction of SCADA (supervisory control and data acquisition) to transmission and distribution subsystem in the late 1970s. The SCADA was introduced to industrial and manufacturing industries in the 1960s to automate the monitoring and control processes associated with their complex operations. Figure 10.2 illustrates typical SONETSDH based power utility network with SCADA systems.

The SONET/SDH ring and other TDM networks such as T1/E1 with 3G/4G connectivity were primarily used for power substations connectivity. SCADA systems were part of such network connectivity for monitoring and control of power distribution system. IRIG-B and NTP were predominant timing solutions for the power substations. Gradually, both network subsystems and SCADA are replaced for greenfield network with increased use of PTP for time synchronization and Ethernet/IP/MPLS for modern smart grid infrastructure.

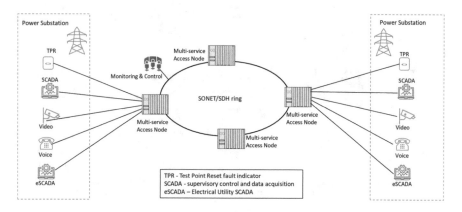

Fig. 10.2 Traditional power utility network with SCADA system

10.3   The Smart Grid

The power grid has been degraded since it's the beginning of its worldwide distribution. The fundamental components include power generation, transmission, distributions, and consumptions. The power generation is done through conventional and modern power generation systems such as hydroelectric generators, combined heat and power (CHP) plants, thermos generators, and nuclear power sources. The degradation of power utilities stemmed from voltage instability, intermittency, curtailments, blackouts, and unbalanced or heavy-load situations. To overcome these issues, several reference models were proposed by NIST (National Institute of Standard and Technology), IEEE, and IEC to allow smart control, visualization, monitoring, and control by integrating physical and cyber communication networks to conventional grid to improve communication and control of power network. Figure 10.3 shows a modern smart grid infrastructure that incorporates various reference models into a comprehensive system.

This comprehensive model of smart grid infrastructure includes overall control and management of power generation through distribution and interconnect of power networks. The interactions at electrical and communication levels are done through a number of communication architectures including area networks, power system architectures of generation, transmission, and distribution, and ICT (Information Communication Technology) architecture as shown in the figure above. The plant control system including power generation transmission and distribution is integrated with other domains through Wide Area Networks (WAN) and local area network such as Neighborhood Area Network (NAN)/Field Area Network (FAN) and Home/Build/Industrial Area Network (HAN/BAN/IAN).

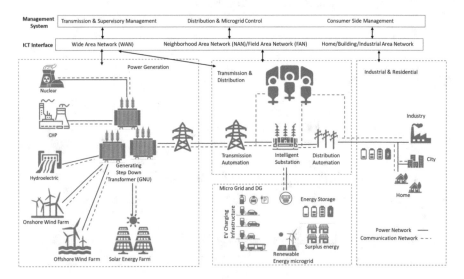

Fig. 10.3 Modern smart grid infrastructure

However, it is to be noted that smart grid implementation is a continual process of improvement in brownfield environment and thus, most power grid upgrade requires careful consideration to the integration of SCADA system, existing network infrastructure and synchronization. Some of the most widely known systems that are included in the brownfield and greenfield development towards a smart grid include RES-Es (electricity from renewable energy sources), DG (Distributed Generation) sources, AMIs (Advance Metering Infrastructures) that are deployed at customer levels, smart measurement systems for active and reactive power detection, eSCADA (electrical SCADA), improved control systems based on wired and wireless communications technologies, and measurement and monitoring devices such as synchrophasors, PMUs (Phase Measurement Units), power quality analyzers (PQAs), and sensor networks.

10.4   The Need for Synchronization

Time synchronization is an important application of power grid, specifically for smart grid. Knowing the state and health of this critical resource is the more important now than ever given the expanse of interconnected network of power system and its span across geographic distances and countries. The large blackout event of August 14, 2003 affecting much of eastern North America is a good example that depicts what happens in one part of the grid which affects operation elsewhere. The actual collapse, a cascade of outages that only took 9 seconds affected 50 million people in eight states and the Province of Ontario. Yet, most experts agreed that the conditions leading up to this event were years in the making with consequences both foreseeable and inevitable. Those conditions included the inability of power operators to: (1) adequately assess the situation before the fact; (2) respond quickly enough once events started to unravel; (3) coordinate their actions in real time across the region; and (4) accurately timeline events after the fact and therefore pinpoint causes and prevent future outages [6]. The consistence casual factor across all these issues is time, specifically time synchronization. It is now well established in the power industry that understanding, and possibly controlling power grid events and their interactions require a means to compare what is happening at one place at a given time with that happening at other places at the same time.

In October 2016, NIST (National Institute of Standard and Technology) conducted a workshop to identify the challenges, the community of experts and potential collaborators as well as key research priorities to guide future NIST efforts in ensuring that the integrity, availability, accuracy, and precision of timing requirements are met in power systems. The workshop identified several timing issues and challenges faced by power grid operators. Utilities and literatures reference during workshop and the literature [5–9] surveyed by this author found following key applications requirements:

1. *Traveling-Wave Fault Detection*: One of the most stringent application identified during the NIST workshop was "Traveling-Wave Fault Detection," which requires synchronization on the order of hundreds of nanoseconds in order to precisely locate a fault to the scale of hundreds of feet as the electricity waves are traveling at near the speed of light.

2. *Synchrophasor*: To isolate the fault and make timely decision, synchrophasor is among the key technologies in building awareness among the grid to make timely decisions on the network. These timestamped power quantities are used for monitoring, control, prediction, and optimization applications. Among the main uses with the most stringent time requirements is the ability to take time-synchronized measurements using PMUs and provide consistent reporting rates at 60 times per second [7].

3. *Wide Area Measurement System (WAMS)*: Frequency event detection that covers hundreds of miles is important as discrepant shifts in phase angles alert system operators to dynamic voltage and frequency stability issues. The higher precision reporting rates provide higher resolutions that can detect fast transients in the system.

4. *Event Timing and Reconstruction*: In the aftermath of the eastern blackout of August 2003, North American Electric Reliability Council (NERC) blackout investigators found that many of the various disturbance recordings were not synchronized, which made blackout analysis work significantly more difficult and time consuming. Event timing and reconstruction ultimately serves the purpose of event analysis to understand what actually happened, when it happened, and why it happened.

5. *End-to-End Relay Testing*: Today, utilities are investing in new and upgraded protective relaying schemes on existing and new transmission lines and equipment as they improve their systems. The new schemes are leveraging new technology to improve the speed of operation, reliability, and data gathering. Other beneficial features are present that were never available before. Precision time synchronization is important in such application.

6. *Multi-rate Billing*: We have perhaps noticed in our utility bill more often than not, different pricing is applied in a tiered pricing structure offering incentives to use power during off-peak time. Precision synchronization is important in such multi-rate billing.

7. *Power Quality Incentives*: Measuring power qualities are important as it allows utilities to identify potential quality impairments. Precision time accuracy is needed for such measurement.

8. *Application to IEC 61850 Processor Bus*: Precision time synchronization in IEC 61850-9-2 process bus is extremely important to enable critical and control function in the substation.

10.4.1   Traveling-Wave Fault Detection (TWFL)

Fast and accurate fault detection by locating both permanent and temporary faults on transmission lines is of great value to power transmission asset owners and operators. Fault locating as a discipline dates back to the 1940s and continues to evolve [8]. These faults in power transmission lines cause transients that travel at a speed close to the speed of light and propagate along the line as traveling waves. Accurately locating faults on power transmission and distribution systems is very important for utilities to allow quick maintenance and repair action of the repair crew. In early days, visual inspection was critical to locate fault in which gun powder cartridge was installed in target towers. As the fault transients traveled to ground through the tower, it would heat up the charge and set the target pointing to the fault location. This approach significantly helped ground patrols identify damaged insulators. Visual inspection has evolved from road helicopter and drones, but it is now used for other purposes including vegetation management, etc. after detecting the fault location. Using electromechanical device conduct to electrical measurements has evolved today to microprocessor-based systems integrated with geospatial data. Advances in technology including high-speed sampling, digital signal processing, precise time synchronization, and digital communications enable further improvements in TW (Traveling Wave) fault locating. The TW fault locating has recently been integrated with microprocessor-based line protection relays improving convenience and reducing cost of ownership to utilities [9, 10].

These TW faults occur at any point on the voltage wave other than at voltage zero launches a step wave, which propagates in both directions from the fault location as shown in Fig. 10.4. One method of determining the fault location is to use precise measures of the TW arrival times at both ends (herein point A and Point B) of the transmission line.

The calculation to locate fault is quite simple. If we know the speed of the traveling wave (its relative difference to the speed of light) and the length of the line being

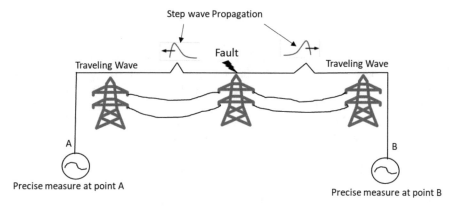

Fig. 10.4 The precise measurement of TW faults

monitored, then we can work out the distance to fault by using the time difference of the arrival times of the traveling waves at each end of the line [11].

Distance from endpoint A = [Line length
                          + (Time endpoint A – Time endpoint B).v] / 2

Distance from endpoint B = [Line length
                          + (Time endpoint B – Time endpoint A).v] / 2

where V is propagation velocity of the traveling wave.

Fault location systems have been traditionally relied on the measurement of power frequency components. However, TW fault location has gained an increasing interest to utilities and power researchers due to its accuracy. Today, fault locator systems measure the time of arrival of a fault-generated traveling wave at the line terminals using the precise timing signals from the GPS. There are three methods of locating TW fault, one of them we just discussed in which time tags the arrival of the fault-generated surges at two time-synchronized locations, usually the ends of the line. The distance is determined in terms of the difference of the arrival times. A second method is to calculate fault distance by analyzing the fault-generated traveling-wave waveforms recorded at one end of the line. The time difference between the initial fault surge and the consecutive reflected pulse is the time interval for a surge to travel from terminal to fault and back. The third method is to make use of transients generated when a circuit breaker is closed to a dead line. The time interval between the pulse created by breaker closing and the reflected pulse from a short circuit, open circuit, or broken conductor is used to calculate the distance to fault. Accuracies of sub-microseconds are required for traveling-wave fault location [8]. Today, PTP-based timing solutions (with hardware timestamping) are capable of achieving 100 ns (0.1 μs) or better accuracies.

10.4.2 Synchrophasor

Synchrophasors are time-synchronized numbers of a quantity described by a phasor that includes magnitude and phase information. These are represented by both the magnitude and phase angle of the sine waves found in electricity and are time-synchronized for accuracy. The synchrophasor technology uses monitoring devices such as phasor measurement unit (PMU) for the high-speed measurements of phase angles, voltage, and frequency that are timestamped with high-precision clocks. The data reporting rates for these measurements by PMUs are typically 30 to 60 records per second, and may be higher; in contrast, SCADA systems often report data every 4–6 s which over a hundred times slower than PMUs.

The accurate time resolution of synchrophasor measurements allows unprecedented visibility into system conditions, including rapid identification of details such as oscillations and voltage instability that cannot be seen from SCADA

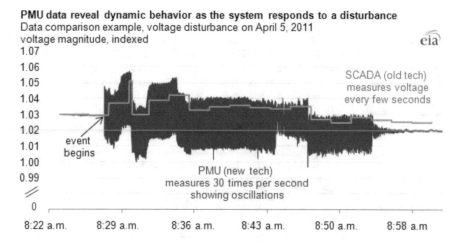

Fig. 10.5 SCADA measurement of voltage compared with PMU measurement of voltage (Courtesy: US Energy Information Administration based on Oklahoma Gas & Electric system disturbance data) [13]

measurements. Complex data networks and sophisticated data analytics and applications convert PMU field data into high-value operational and planning information [12]. Figure 10.5 depicts comparative measurements of SCADA and PMU plotted for voltage disturbance on April 5, 2011 presented by eia [13] in a blogspot. The data shows unprecedent visibility into system conditions, including rapid identification of details such as oscillations and voltage instability that cannot be seen from SCADA measurements.

The SCADA system was unable to cope with the data reporting rates of high than 4–6 s that is required to capture power system dynamics and thus as shown in the diagram it did not capture the accurate picture of oscillation that has occurred. Ideally, we need data at sub-second rates these days in order to adequately monitor power system dynamics.

In case of 2003 blackout in the USA and Canada, if PMUs were deployed at the time, the issue of blackout could have been prevented. For example, the bus voltage angular separation presented in the figure would have been detected by the PMU. To understand the conditions of power system measurement of different power qualities are important, e.g., current and voltage. Another important element that can provide a better insight to anomalies in power system is angle. In an AC power system, power flows from a higher voltage phase angle to a lower voltage phase angle: the larger the phase angle difference between the source and the sink, the greater the power flow between those points implying larger the static stress being exerted across that interface and closer the proximity to instability [14]. It has been seen that during stress angular separation between bus voltages increase as evident in the diagram. During the 2003 blackout, angular separation (please refer to Fig. 10.6) started to increase as indicated by red and blue line and just before the blackout it was almost 45° and during the blackout nearly 115°. If that point of time,

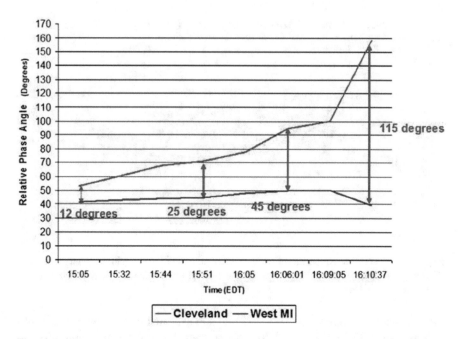

Fig. 10.6 Bus voltage angular separation during the 2003 blackout (Courtesy: CERTS) [14]

PMUs were part of the network, operator could have seen the stress in the system, and preventive action could have been taken.

Today, complex data networks and sophisticated data analytics and applications allow conversion of PMU field data into high-value operational and planning information.

10.4.2.1 Phasor Measurement Unit (PMU)

The PMU is a type of synchrophasor technology that enables operator to measure electrical quantities such as current voltage and the values of power frequency and rate of change of frequency (ROCOF) at diverse locations in a power grid. The measured data output is accurately timestamped and transmitted to a Phasor Data Concentrator (PDC) central analysis station in accordance with the standardized transmission protocol IEEE C37.118. The PDC in turn provides the data to other applications for visualization and analysis. Figure 10.7 shows multiple PMUs connected to local PDC which is in turn connected to central PDC where operator can store data and visualize it for analysis using application user interface. The IEEE C37.244-2013 PDC Guide provides description to different functions of PDC as listed below:

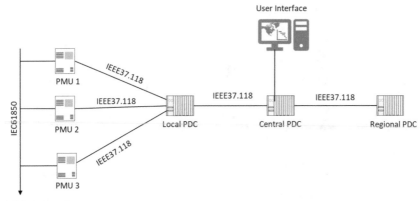

Fig. 10.7 PMU and PDC connectivity

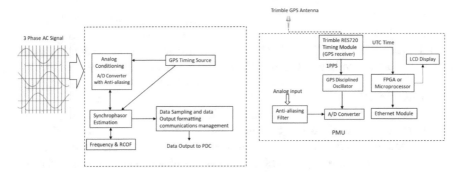

Fig. 10.8 Interworks of PMU along with PMU block diagram

- Data aggregation, forwarding, validation, and communications.
- Data transfer protocol support and protocols conversion.
- Data format and coordinate conversion.
- Data latency calculation, reporting rate conversion, and output data buffering.
- Configuration and performance monitoring.
- Phase and magnitude adjustment.
- Redundant and duplicate data handling.
- Data retransmission request and cyber security.

While PDCs collect data for various purposes as described above, PMUs are critical device in that it collects data directly from status bus in a given substation and feeds that data to PDCs for analysis and visualization. The PMUs can be stand-alone or integrated in another device. Today, many PMUs also support PTP for network wide time synchronization in addition to PRTC input directly from an integrated GPS timing module. Figure 10.8 shows the interworks of PMU along with the block diagram of PMU.

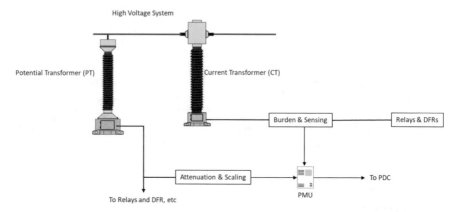

Fig. 10.9 Typical PMU deployment in substation for the measurement using PT and CT sensing [15]

A typical PMU consists of A/D frontend that interfaces analog signals, a time input synchronizing the measurement, and a processing section that makes the phasor, frequency, and ROCOF estimation. The processor unit manages output formatting and data transmission. Figure 10.9 illustrates a typical deployment of PMU in substation for the measurement using Potential Transformer (PT) and Current Transformer (CT).

To properly isolate issues in a substation, consideration must be given to the measurement chain of PMU which may start at transducer, e.g., PT and CT. The second and most important element is measurement of phase angle for which calculation is based on a time signal. There is a direct correlation between time and phase angle error. If the time error is larger, angle error would larger as well. The timing error includes PMU internal process synchronization, error synchronizing with input signal and its synchronization with the reference. Grid-wide angle comparison requires end-to-end synchronization for the grid [15]. While there remain chances of system-induced errors, PMU does better in comparison to older SCADA system as the latter has many shortcomings including its inability to report phase angle of voltage and current signals. The synchrophasor technology such as PMU is a popular choice and widely deployed in smart grid.

10.4.3   Wide Area Measurement System (WAMS)

Today, increased electric energy consumption and power system restructuring have posed new challenges to the operation, control, and monitoring of power systems. The SCADA system is simply not good enough to ensure the security and stability of modern power system. Moreover, SCADA is often unable to measure data of all buses simultaneously. Additionally, as discussed earlier, SCADA is unable to

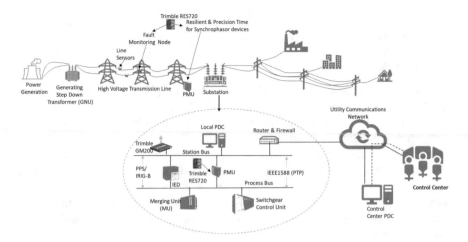

Fig. 10.10   Typical WAMS setup

adequately report power system dynamics. To overcome this situation, wide area measurement system (WAMS) is increasingly deployed in many power grids. The WAMS consists of advanced measurement technology (such as PMU), information tools, and operational infrastructure that facilitate the understanding and management of the increasingly complex behavior exhibited by large power systems [16]. In certain deployment, WAMS is used as standalone infrastructure complementary to conventional SCADA. In such deployments, WAMS enhances the operator's real-time "situational awareness" that is necessary for safe and reliable grid operation in addition to supporting post-event analysis of significant system disturbances. Today, WAMS technologies are incrementally incorporated into the actual control system of the grid.

The main idea of WAMS is to create centralized data analysis capabilities in which data are collected from different parts of the power grid simultaneously and synchronized with each other. The purpose of analyzing this data is to evaluate the actual operating conditions of the system at any time and to compare the network parameters (voltage and current values and angles, temperature, active and reactive power) with the standard or predetermined limits. Also, with the help of certain algorithms, the security margin of stability or the distance of the system from the stability boundary is determined. Figure 10.10 depicts typical WAMS setup with time synchronization devices and synchrophasor technologies such as PMUs.

Normally, WAMS consists of three processes: Data measuring and collecting, data delivering and communication, and data analysis. The diagram above shows various infrastructure components of WAMS that includes [17]:

- *Synchrophasor devices* such as PMU that must be installed at the key points of the network.
- *Synchronizing system* that provides overall time synchronization for the network.

- *Communications system and infrastructure* that provides appropriate speed, reliability, and security for data transfer.
- *Data collection and analysis center*, which should be equipped with appropriate software for data analysis.

PMU serves as the mainstay in WAMS. At the WAMS control center, data from all PMUs of the network are collected to provide dynamic and observable studies of the system that can be performed in real time. As stated earlier, WAMS can also be integrated with SCADA depending upon deployment scenario to support brownfield upgrades. In addition, WAMS also assists in the protection system configuration by using wide area data including remote buses and current transformers conditions indicating that these devices are not saturated and do not suffer from voltage fluctuations. Today, WAMS requires stringent time synchronization to adequately collect and present power dynamics. Table 10.1 depicts time synchronization requirement for WAMS.

10.4.4   Event Timing and Reconstruction

Power system no doubt has become marvel of reliability and in fact reliability and maintainability are two hallmarks of power system. In their most advanced form, defined in the Uptime Institute's Tier IV classification, they have "system + system" (S + S) redundancy, 2 active power delivery paths, and must be fault-tolerant and concurrently maintainable. However, reliability principle puts a limiting factor to such aspiration in that no power system can operate 100% of the time indefinitely due to several factors including component degradation. Even Tier IV systems, as well-designed as they are, are subject to this statement, as evidenced by representative site availability figures which are less than 100% [18]. Backup and recovery procedures are used to save the critical load in an emergency condition. However, the key aspect of incident recovery is the ability to quickly understand what went

Table 10.1 Wide area precision time requirements in current power systems [7]

Application	Time accuracy requirements
Traveling-wave fault detection and location	100–500 ns
Synchrophasors, wide area protection, frequency event detection, anti-islanding, droop control, and wide area power oscillation damping (WAPOD)	Better than 1 μs
Line differential relays	10–20 μs
Sequence of events recording	50 μs to ms
Digital fault recorder	1 ms
Communication events	
Substation local area networks (IEC 61850 GOOSE)	100 μs to 1 ms
Substation local area networks (IEC 61850 sample values)	1 μs

Fig. 10.11 Cascading events and state of power system

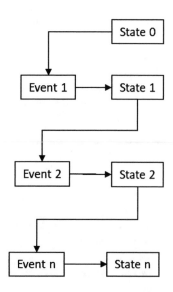

wrong and take corrective actions. Assuming that the system is professionally designed and constructed, incidents do happen and generally stem from component failure due to internal or external factors. Therefore, objectives should be to mitigate incidents by determining what happened and how to prevent this from occurring again.

For this purpose, we can categorize incidents into a series of events to better analyze them and take preventive measures. These events may will eventually stop, leaving the system in either an operating, partially operating, or non-operating condition as depicted in Fig. 10.11.

To analyze this the question should what caused first event. Obviously, the causal factor can be internal or external elements. To understand what went wrong, series of events should be measured, and their effects are analyzed through instrumentation as much of their manifestations are not measurable through human eye. The final and more important is the complication of measured events in order of occurrence via a record of the times at which they occurred. Therefore, it is necessary to reconstruct a series of events in an electrical system via recorded electrical measurements. Additionally, the measuring instrument should be fast enough to record these measurements in terms of when they happened. Once the measurements are ordered, the events themselves may be reconstructed to analyze the root cause of the incident. For this matter time is crucial factor. High-precision synchronization within a power system helps sequencing event measurements, scheduling outputs, synchronizing actuation, timestamping logged data, and coordinating events.

Table 10.2 depicts representative timeframe for various electrical events [18].

Please note, the time frames provided here for electrical events are representative example. Readership should consult with manufacturer specification of the system since system-related phenomena timeframes are dependent upon the parameters of the system.

Table 10.2 Representative timeframe for various electrical events [18]

Event	Representative timeframe
60 Hz power system cycle	16.667 ms
Third-harmonic (180 Hz) power system cycle	5.556 ms
Fifth-harmonic (300 Hz) power system cycle	3.333 ms
Seventh-harmonic (420 Hz) power system cycle	2.381 ms
Voltage sag	8.333 ms–1 min
Lightning strike (return stroke)	42 μs
Low-voltage circuit breaker, instantaneous trip from initiation of overcurrent	44 ms
Low-voltage circuit breaker, short-time trip from initiation of overcurrent	200 ms
Overcurrent relay instantaneous trip	35 ms
Bus differential relay trip	6 ms
Lockout relay trip	8 ms
Low-voltage circuit breaker, time from trip to fault clearing (arcing time)	16 ms
Medium-voltage circuit breaker (3-cycle), time from trip to fault clearing (arcing time)	32 ms

As evident from the example of the time frame for representative electrical events, highly accurate time synchronization is needed for measured events and analysis. Without time synchronization, it can be difficult or impossible to correlate the time sequence of events involving multiple network components. Providing accurate event timestamps can eliminate ambiguity and maximize analysis efficiency when reconstructing major events such as cascading faults and blackouts. Typically, 1 ms may be needed for synchronization of DME (Disturbance Measuring Equipment) devices such as sequence of event recorders, digital fault recorders, and relays with event recording functions [8].

10.4.5 End-to-End Relay Testing

Precision time synchronization is required to coordinate the actions of protective relays, generator controllers, distributed resources, power system controllers, energy storage, and microgrids. Every device must be synchronized to the same time source to ensure coordinated action. The protective relay is one of the most important devices for transmission line protection and has been deployed over a century now to protect transmission line by isolating defective elements. These devices detect the fault and initiate the operation of the circuit breaker to isolate the defective element from the rest of the system. Over the last 60 years, protective relays have undergone considerable changes, most obvious of which is reduction of its size. Today, these devices are compact and self-contained and used to detect the abnormal conditions in the electrical circuits by constantly measuring the electrical quantities which are different under normal and fault conditions. The electrical quantities (voltage, current, frequency, phase angle, etc.) which change during fault conditions present the type and location of the fault to protective relays. Having detected such condition, the relay closes the trip circuit of the breaker resulting disconnection of the faulty circuit. Figure 10.12 illustrates a typical protective relay.

Today, many protective relays are available with integrated timing modules and support PTP. Figure 10.13 shows typical PTP-based time synchronization in a substation that includes protection relays.

The protective relaying minimizes the damage to power equipment and interruptions to the service during an electrical failure. Such scheme includes protective current transformers, voltage transformers, protective relays, time delay relays, auxiliary relays, secondary circuits, trip circuits, etc. Apart from substation power

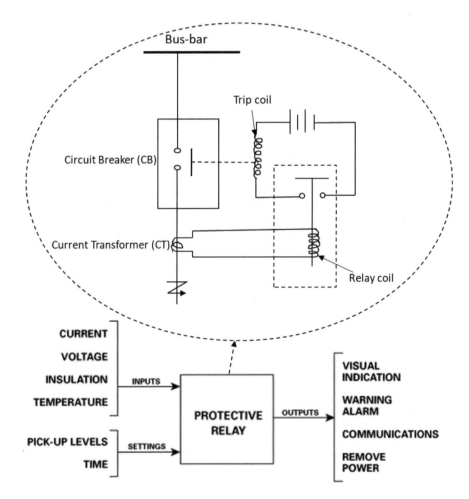

Fig. 10.12 A diagrammatical representation of protective relay

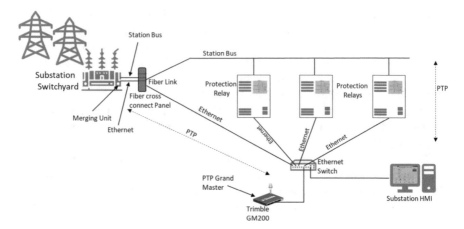

Fig. 10.13 Substation network with protection relays and PTP-based time synchronization

system, protective relays are also deployed for transmission line protection. The protective relay scheme deployed to protect a transmission line will require the proper coordination of relay settings at the two ends of the line. Engineers use models of the power system to determine these settings and then verify the proper operation of the relays using two relay test sets, each connected to a relay at the two ends of the line. For this purposes, end-to-end testing is considered as the ultimate test for protective relay protection schemes with two or more relays communicate trip and blocking information with each other. End-to-end testing can provide the most realistic fault simulations to prove relay protection schemes before placing them into service and this test technique is becoming more and more popular, especially as the National Electrical Reliability Council (NERC) and other regulatory agency standards are becoming more stringent. Figure 10.14 depicts typical setup for end-to-end testing. Protective relays are connected to relay test equipment which is getting timing input from Trimble thunderbolt disciplined clock or GNSSDO (GNSS disciplined oscillator). The GNSSDO provides highly accurate timing output that is essential for end-to-end testing. The protective relays are connected to the transmission line through a T-line junction point for end of the line measurement testing. Engineers at each end communicate, select the appropriate test, enter the test time and then initiate the test.

In this setup, test run automatically for which test phasor values and test time are loaded to the test instrument. The time decoded from the GNSSDO and the test time from the program are continually compared. When the receiver time equals the test time, the test data are replayed synchronously at each end of the line [19]. As evident from this test setup precise time synchronization plays an important role in protective relay testing.

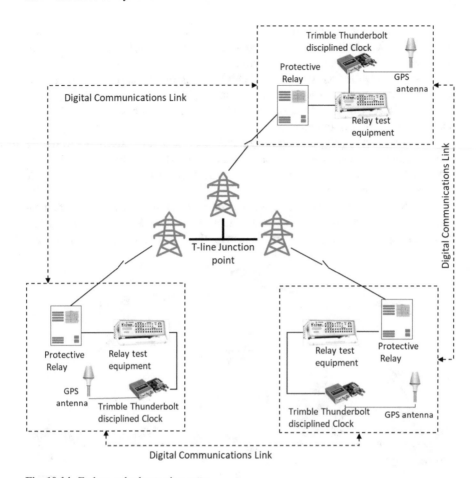

Fig. 10.14 End-to-end relay testing setup

10.4.6   Multi-Rate Billing

To improve overall system utilization, utilities offer certain customers incentives to use power at off-peak hours by providing them with lower energy rates during these periods. While this seems simple enough, the amounts of money can be considerable, and a meter with a simple real-time clock that slowly drifts off (until it is reset) probably will not be acceptable as a basis for determining billing periods. Precise and economical timing solutions now make it possible to provide highly accurate revenue metering technology at the customer level, providing potential savings. Providing accurate, traceable time to both operator and customer eliminates the situation where different measurement intervals yield different results, possibly leading to conflict between utility and customer. Accuracies in the order of milliseconds are used here. One of the critical technologies used for multi-rate billing is AMI (Advanced Metering Infrastructure) which is an integrated system of smart meters, communications networks, and data management systems that enable two-way communication between utilities and customers. The system provides numerous important functions that were previously not possible or done manually. These functions include the ability to automatically and remotely measure electricity use, connect and disconnect service, detect tampering, identify and isolate outages, and monitor voltage. Combined with customer technologies, such as in-home displays and programmable communicating thermostats, AMI also enables utilities to offer new time-based rate programs and incentives that encourage customers to reduce peak demand and manage energy consumption and costs [20]. Figure 10.15 shows how AMI and customer system works to collect usage information and provide customer with multi-rate billing.

The AMI deployment includes three key components, smart meters, communications networks, and a meter data management system (MDMS). The AMI utilizes customer system to help customer manage their electricity consumption and

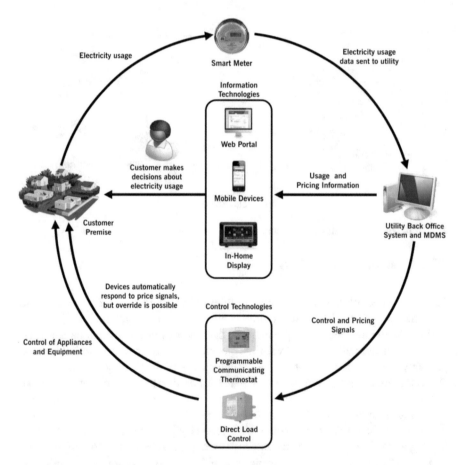

Fig. 10.15 A diagrammatical representation of how AMI and customer system works to collect usage information and provide multi-rate billing (courtesy: energy.gov) [20]

associated cost. The customer system includes programmable communicating thermostats (PCT) and direct load control system that both utilities and customer use automatically control customer's heating and cooling system or other energy intensive devices. Additionally, HAN (home area network) and energy management system can be installed to automatically control price signals and load conditions.

Electricity consumptions can be presented to customers through in-home display (IHD), web portal, text messages, and emails. To ensure accountability, time synchronization across AMI network as well as accurate timestamp of collected data is vital.

10.4.7   Power Quality Incentives

Utilities and system operators are driven to maintain power quality as much as possible and now considering financial incentives to customers to help maintain power quality. Certain types of load are known for causing power quality impairments, such as harmonics and flicker. Harmonics are a frequent cause of power quality problems. It can be generated when a non-linear load, such as a rectifier, is connected to the system. Random or repetitive variations in the RMS voltage between 90% and 110% of nominal can produce a phenomenon known as "flicker" in lighting equipment. Abrupt, very brief increases in voltage, (called "spikes," "impulses," or "surges"), are generally caused by large inductive loads being turned off [8].

These impairments may occur at customer installation and propagate onto power grid and affect other customers. The power quality impairments are measured in accordance with international standards, such as the IEC 61000-4 series and summarized in reporting periods, generally 10 min. As with multi-rate billing, the amounts of money can be significant, so customers often conduct their own measurement in parallel with that of the utility. Hence, it is important to provide primary reference time traceable to both customer and operator to eliminate disagreement if any. Normally, an accuracy of 1 ms is used for measuring to synchronize periodic measurements of power-quality quantities such as harmonics and flicker.

10.4.8   Applications to IEC 61850 Processor Bus

Communications play critical role in real-time operation of power system. At early days, telephone was used to communicate "line loadings" back to the control center as well as to dispatch operators to perform switching operations at substations. In the 1930s, telephone-switched remote control units were used to provide status and control for a few points. Later in the 1960s data acquisition systems (DAS) were installed to automatically collect measurement data from the substations. Due to limitation of bandwidth at the time, DAS communication protocols were optimized to operate over low-bandwidth communication channels. As high-speed digital communications become a common place, IED (Intelligent Electronic Device) receives and communicate thousands of analog and digital data points and for this bandwidth is no longer the limiting factor. As IEDs are increasingly deployed in substations to perform electrical protection functions and advanced local control intelligence, data becomes critical for substation control. A key component to this communications system is the ability of the devices to describe themselves in both data and services that includes multi-vendor interoperability, high-speed IED to IED standard based communication, autoconfiguration and support for voltage and current sample data. Given this requirement, a communication architecture was developed in 1988 by UCA (Utility Communication Architecture) that intended as a framework to address and help resolve interoperability issues among devices and create seamless communications path and integration. The result was a set of recommended protocols that adheres to various layers of OSI reference model for communications. The concept and the works done at UCA framework serve the basis for IEC 61850 standard. The IEC 61850 is de facto international standard for substation automation. It allows the merging of communications capabilities of all IEDs in a substation or even an entire power network for data and control to ensure more efficient protection, monitoring, automation, metering and control functions. The standard defines communications between devices in the substation and the related system requirements. It supports all substation automation functions and their engineering. The ideas behind IEC 61850 are also applicable in areas of automation such as control and monitoring of distributed generation [8]. Figure 10.16 shows IEC 61850 communication protocol stack (Fig. 10.16a) and substation automation architecture (Fig. 10.16b).

IEC 61850 uses object-oriented methods for logical communications between substations, primary process equipment, and secondary devices. Using this object-oriented model, IEC 61850 extracts the information templates from the communication details. The protocol implements a producer/consumer model by incorporating Quality of Service (QoS) and multicast to allow any unit to communicate with other units. It also takes advantage of the Manufacturing Message Specification (MMS), an international standard to transfer real-time process data and supervisory control

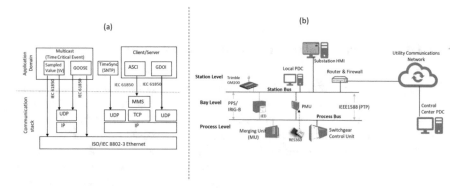

Fig. 10.16 IEC 61850 communication protocol stack and substation architecture

information between networked devices and/or computer applications. IEC61850 uses Ethernet as underlying communications medium. To support next generation substation architecture, the ability to digitize the base power quantities at the source and transmit the resulting sample values back to the substation becomes a need. In addition to Sampled Values (SV), the ability to remotely acquire status information as well as set output controls is very desirable [21]. The IEC 61850 standard addresses this need through the definition of "Sampled Measured Values" services and the implementation of a process bus. The figure above depicts IEC61850 architecture for substation automation that allows collection of various power quantities through defined communication paths example station bus and process bus. The station bus interconnects all bays (please refer to figure above) with the station supervisory level and carries control functions while process bus connects the IEDs and merging units within a bay and carries real-time measurements data such as SVs. Unlike station bus, the process bus provides digital link to primary equipment like switchgear and instrument transformers. The IEC 61850 process bus is defined by IEC 61850-9-2 process bus standard. It reduces conventional copper wiring and TCO (total cost of ownership) by allowing power quantities as well as control commands to switchgear in the switchyard to be distributed over an Ethernet network for connection to protection and control IEDs. Today, many IEDs and MUs support packet-based time synchronizations such as PTP; however, the entire process bus requires stringent time synchronization of sub 1 µs accuracy.

10.5 PTP Power Profile

Thus far we discussed about various synchronization needs for power system including that of digital substation. For the purpose of building a digital substation and precision time-synchronized power networks, PTP is essential as it provides highly accurate time synchronization unlike NTP. To operate PTP in a power network, power profile should be configured. The profile options are: IEC 61850-9-3, IEEE C37.238-2011, and IEEE C37.238–2017. IEC 61850-9-3 defines a PTP Utility Profile for use in substation applications while IEEE C37.238-2017 builds upon IEC 61850-9-3 and adds few additional parameters.

The IEEE/IEC 61850-9-3 base power profile includes the following [22]:

- 1 s intervals for PTP messages
- Multicast communications and Layer 2 mapping
- Peer-to-peer path delay measurement
- Steady-state performance requirements
- One-step and two-step clocks
- Default best master clock algorithm
- Local time type length value (TLV) extensions

Parameters specific to the IEEE C37.238-2017 extension include the following [22]:

- Pdelay is optional for slave-only clocks (for when the delay of the end cable is insignificant)
- IEEE_C37_238 TLV
- IEEE 1588 alternate time offset indicator (ATOI) TLV
- Mappings for IEEE C37.118 and IEC 61850 protocols (informative)
- IRIG-B replacement mode (by tightening the ATOI specifications)
- Real-time adjustments of the advertised time-quality (so IEDS can determine if acceptable)

To be compliant with IEC 61850-9-3 standard, a device must agree to at least one of the following three management mechanisms:

- SNMP MIB in accordance to IEC 62439-3:2015
- Management objects defined in IEC 61850-90-4:2013
- Manufacturer-specific implementation to address all configurable values stated in IEEE 1588-2008 clause 15.1.1

Furthermore, IEC 61850-9-3 also defined PTP attributes in Table 10.3. The PTP attributes are chosen in a way that clocks compliant to the default peer-to-peer profile of IEEE 1588-2008 can be configured to lock to an IEC 61850-9-3 grandmaster clock.

Additionally, IEC 61850-9-3 standard also specifies overall network time budget less than 1 µs for which a configuration of max 15 transparent clocks and 3 boundary clocks can be done. Table 10.4 shows accuracy needed for the involved clocks:

Table 10.3 Important PTP attributes of IEC61850-9-3 power profile

PTP attribute	Value
Domain number (defaultDS.domainnumber)	0
Log announce interval (portDS.logAnnounceInterval)	0
Log sync interval (portDS.logSyncInterval)	0
Log min delay request interval (portDS.logMinPdelay_ReqInterval)	0
Announce receipt timeout (portDS. announceReceiptTimeout)	3

Table 10.4 Clock accuracy required as per IEC61850-9-3

Clock type	Accuracy needed
Grandmaster clock	±250 ns
Transparent clock	±50 ns
Boundary clock	±200 ns

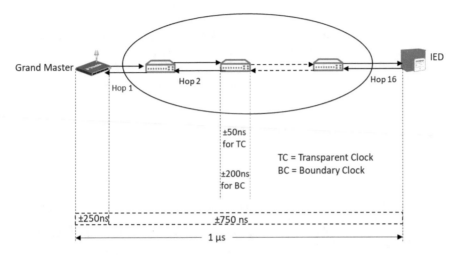

Fig. 10.17 Total end-to-end network time error budget for substation network as per IEC 61850-9-3

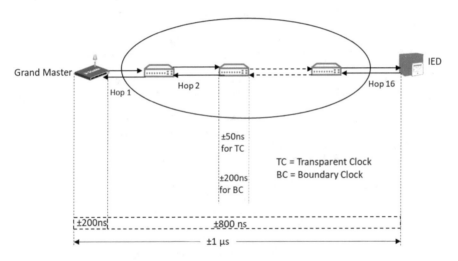

Fig. 10.18 End-to-end network time error budget as per IEEE C37.238-2017

If we apply the involved clocks accuracy requirements to a typical network setup, the total end-to-end network time error budget should be less than 1 μs as depicted in Fig. 10.17.

Similarly, IEEE C37.238–2017 also specifies that end-to-end network time error budget is less than 1 μs. As per the standard, a maximum of 16 hops is accepted and the time reference source can contribute up to 0.25 μs to the total time error, BCs can contribute up to 0.2 μs, and network devices (transparent clocks) can contribute up to 50 ns each. Figure 10.18 shows end-to-end time error budget as per IEEE C37.238-2017.

Please note, the standard suggested that a grandmaster can introduce a maximum of 200 ns and the entire network should be within 800 ns. The steady-state performance is considered for network loads at 80% of the wire-speed. Additionally, to support quality of time distribution service during grandmaster change, IEEE Std C37.238 specifies holdover drift for grandmaster-capable devices to be within 2 μs for up to 5 S at a constant temperature.

References

1. Edisontechcenter.org. (2014). *The history of electrification*. Edison Tech Center.
2. Simões, et al. (2014). *Comparison of smart grid technologies and progress in the USA and Europe*. Green Energy and Technology.
3. Amin, M. (2004). Balancing market priorities with security issues. *IEEE Power Energy Magazine*, 1540–7977/04.
4. Amin, M. S., & Wollenberg, F. B. (2005). Toward a smart grid. *IEEE Power Energy Magazine* 1540–7977/05.
5. USDE. (2018). *Smart grid system report. 2018 Report to congress*. US Department of Energy.
6. Symmetricom. (2004). *How time finally caught up with the Power Grid: Whitepaper*. Symmetricom, Inc.
7. Allnutt, J., Anand, D., Arnold, D., Goldstein, A., Li-Baboud, Y., Martin, A., et al. (2017). *Timing challenges in the smart grid*. NIST Special Publication 1500-08. National Institute of Standards and Technology.
8. Aweya, J., & Al Sindi, N. (2013). Role of time synchronization in power system automation and smart grids. In *2013 IEEE international conference on industrial technology (ICIT)*. IEEE Xplore. 23 April 2013.
9. Schweitzer, O. E., Guzmán, A., Mynam, V. M., Skendzic, V., & Kasztenny, B. (2014). Locating faults by the traveling waves they launch. In *2014 67th annual conference for protective relay engineers*. IEEE Xplore: 24 April 2014.
10. Marx, S., Johnson, B. K., Guzmán, A., Skendzic, V., & Mynam, V. M. (2013). Traveling wave fault location in protective relays: design, testing, and results. In *Proceedings of the 16th annual Georgia tech fault and disturbance analysis conference*, Atlanta, GA, May 2013.
11. QUALITROL. (2020). *What is traveling wave fault location? An introduction to traveling waves in a power system or transmission lines*. Qualitrol Company LLC.
12. NASPI. (2014). *Synchrophasor technology fact sheet*. North American SynchroPhasor Initiative.
13. eia. (2012). *New technology can improve electric power system efficiency and reliability*. US Energy Information Administration.
14. CERTS. (n.d.). *Why are phase angle differences important? – Advance concept FAQ*. Consortium for electric reliability technology solution. Electric Power Group. Available online at http://www.phasor-rtdms.com/phaserconcepts/phasor_adv_faq.html
15. Nuthalapati, S. (2019). *Power system grid operation using synchrophasor technology*. International Publishing AG, part of Springer Nature.
16. Hadley, D. M., McBride, B. J., Edgar, W. T., O'Neil, R. L., & Johnson, D. R. (2007). *Securing wide area measurement systems*. Pacific Northwest National Laboratory operated by Battelle for the United States Department Of Energy under Contract DE-AC05-76RL01830.
17. Alhelou, H. H., Abdelaziz, Y. A., & Siano, P. (2021). *Wide area power systems stability, protection, and security*. Springer.
18. Brown, B., & Kozlowski, M. (2006). *Power system event reconstruction technologies for modern data centers*. Schneider Electric.

19. Havelka, J., Malarić, R., & Frlan, K. (2012). Staged-fault testing of distance protection relay settings. *Measurement Science Review, 12*(3), 111–120.
20. Smartgrid.gov. (2016). *Advanced metering infrastructure and customer system. Results from the smart grid investment grant program.* US Department of Energy. Office of the Electricity Delivery and Energy Reliability.
21. Baigent, M. A., & Mackiewicz, R. (2009). *IEC 61850 communication networks and systems in substations: An overview for users.* GE Digital Energy.
22. IEEE Std 2030.101. (2018). *IEEE guide for designing a time synchronization system for power substations.* IEEE Std 2030.101™-2018. IEEE Power and Energy Society.

Chapter 11
Synchronization for Data Center and MSO Infrastructure

11.1 Introduction

The time synchronization is essential to distributed system and for that matter it is inherent part of data center infrastructure. The difference is that many data center designers are oblivious of it due to the inherent support of OS (operating System) for NTP. The protocol is capable of synchronizing over the internet by connecting to public NTP stratum 1 servers to synchronize local servers and systems if there is no primary server present in the data center. In most cases, data centers do implement local NTP servers and create stratum hierarchy level for clock synchronization. Though NTP remains the de facto time sync protocol in data center, distributed databases, live VM migration, real-time big data, and many industry-specific applications require better precision accuracy in terms of microseconds than millisecond that NTP provides. This requirement for better and accurate time synchronization has already pushed leading cloud service providers to think differently. Let's take Google for example, it addressed the issue of concurrency control for DDBMS with TrueTime synchronization API extended this to global POP (Point Of Presence). Facebook recently augmented their NTP service by replacing ntpd with Chrony to provide better accuracy throughout its data centers. Creating such solutions requires development efforts, tests, and measurements to ensure solutions fit to specific application needs. It is easier said than done; while such massive scale undertakings fit well with hyperscalers such solution is not easy for small- to mega-scale data centers to deploy. Henceforth, lack of a guaranteed high accurate clock distribution network, data centers could create localized PRTC and distributed time sync plane for application-centric PoD. It is a common practice to localize traffic in data center due to today's east to west traffic pattern and this application-centric PoD is necessary for modern data center design. This fits well with localized PRTC and PTP based application-centric sync plane which is extensible with overlay tunnels supported by underlying networking technologies. Similar to data center,

D. D. Chowdhury, *NextGen Network Synchronization*, https://doi.org/10.1007/978-3-030-71179-5_11

synchronization (frequency and/or phase) is a must for cable networks operated by MSOs. The challenges have been for MSOs to scale out their network services and modernized the infrastructure and at the same time create distributed synchronization that is inherent to its technologies. Fortunately, a series of specifications created by Cablelabs pave the way for MSOs to reimagine the future of cable networks. This chapter explores synchronization need for both data center and MSO with guidance and design ideas for future infrastructures.

11.2   Synchronization Need for Data Center

To understand the need for synchronization in data center, let us start with the fundamentals that clock or time synchronization is essential in distributed system and for that matter data center that harnesses distributed system is not different. The distributed system can be understood as a collection of computers connected by high-speed communications network. In such settings, each node can share their resources with other nodes on demand. Hence, there is the need of proper allocation of resources to preserve the state of resources and help coordinate between the several processes. To resolve such conflicts, synchronization is used in which clocks of the communicating nodes should agree upon a common time value. If the system is working on a real-time application, then the clocks should match with traceable primary reference clock, e.g., UTC. That being said, today's data center is a classic example of myriad of complexities when it comes to distributed system. With move towards multi-cloud and hybrid cloud scenarios, synchronization of distributed databases and applications among different infrastructure are getting more complex. Additionally, trends towards distributed data centers to accommodate real-time traffic and processing needs at the edge for various industry verticals spew another layer of complexity for synchronization. Depending upon the services that rendered through standalone or distributed data centers, demand for synchronization varies. Traditional implementation of NTP for synchronization may not suffice myriad of applications and services and careful design of data center time synchronization is important. Let's consider distributed database, live VM migration, and Hadoop Cluster as the example to understand the synchronization needs in data center. All three systems require precision time synchronization, and these systems are essential functions of modern data center. It is to be noted that data centers are using NTP for time synchronization of all systems since long, however, these examples depict the need for accurate clock sync in a data center as clock drift may hinder performance and render in application failures.

11.2.1   Distributed Database Synchronization

The decentralization of organizational units and many other factors including myriad of applications led to database distribution in order to solve data island, high availability, and performance issues. The bottom-up approach on a distributed databases design begins with an existing environment including "data island" to a unified distributed database system (DDBMS). A vital feature of this unification and continuous operation of DDBMS is the synchronization process among the different copies of an entity.

The purpose of developing database is to allow multiple users having concurrent access to database and this is no different for DDBMS. These user accesses could occur for a single server or in the case of replicas of databases in multiple servers. The concurrency control is applied to manage this simultaneous execution of transactions. It basically implies synchronization of concurrent transactions and guarantees that database is always in a consistent state by ensuring a specific access order for data items.

Synchronization is performed in one of two ways either incremental synchronization or complete synchronization. The latter requires complete update of database and transmitted data is updated synchronously. This is relatively simple but requires huge overhead for the network. On the other hand, the incremental synchronization requires much less bandwidth, but updates are complex requiring various methods of incremental data collection to be deployed that includes trigger, timestamp, scanning, etc. For either of these synchronizations, all systems involved must have accurate timing to encourage synchronization. This brings us to fundamentals of time synchronization allowing primary traceable time for all systems in a DDBMS to locked in and adjust their clocks.

11.2.1.1   Audit Log

One method of applying synchronization in database is audit log which is different in each database. The audit log keeps track of all activities that includes who accessed the database, what kind of activities were performed, and what data has been changed. Each database has different audit log and audit log table to be used by audit log. For synchronization, audit log, id, and last timestamp are necessary to record and start synchronization. The timestamp is used to check whether there is new data, changed data, or deleted data from the table at synchronized database according to id record and timestamp, also to make sure if there is overhaul towards the entire data in audit table which certainly will take a long time [1].

Table 11.1 shows typical audit log table in which cost of the item in RS is changing according to time_stamp and is recorded. Therefore, one can easily retrieve the latest cost.

Table 11.1 Typical audit log table [1]

ID	Timestamp	Cost in RS
01	2020-5-1 10:18:50	105
02	2020-5-1 10:18:40	104
03	2020-4-15 6:50:00	106
04	2020-3-16 5:51:05	103
05	2020-2-2 8:16:56	102

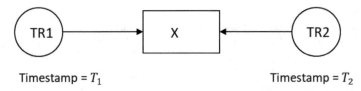

Timestamp = T_1 Timestamp = T_2

Fig. 11.1 Timestamp-based concurrency control

11.2.1.2   Concurrency Control

As stated earlier, concurrency control is used to resolve conflict with the concurrent accessing or changing of data. There are two types of concurrency control, one is lock protocol based and the other is timestamp based. For the former, various lock protocol based mechanisms are used for read and write operations in which a transaction needs to acquire a lock and releases it after the given operation. While this technique serves best for concurrency control, it may lead to deadlocks which will then require deadlock detection, prevention, and avoidance processes [1].

Such issues can be avoided using "timestamp" based protocols in which a timestamp is associated with each transaction as depicted in Fig. 11.1.

Let's assume there are two transactions TR1 and TR2. As shown in the figure, initial timestamp of variable "X" is Timestamp = T_0. When transaction TR1 access variable "X" timestamp is timestamp = T_1 and similarly, timestamp for transaction TR2 is timestamp = T_1. As demonstrated by this example, in a "timestamp" based protocol timestamp of transactions is compared and then access is granted.

The bottom line is that these methods of synchronization require real-time updates and processing of events at microseconds level that is nearly impossible for most NTP deployments.

Some cloud-based DDBMS such as Google spanner and Amazon AWS have addressed this by distributed clock service traceable to primary reference clock. For example, Google TrueTime uses a combination of GPS and atomic clock to guarantee the correct time with precision accuracy. The TrueTime is used by all applications on Google servers, specially Google's "Spanner" DDBMS uses it to assign timestamps to transactions. Such consistency allows cloud base spanner to perform consistent reads across an entire database in multiple regions without write blocking an inherent problem associated with concurrent access.

While Google and Amazon have addressed the need for precision time synchronization in their data centers and across their cloud services, it remains a critical issue for many data centers and cloud services providers. Moreover, a number of DDBMS such as CockroachDB and TiDB depend on tightly synchronized clock by design. Therefore, deploying a distributed precision time synchronized network in data centers is becoming increasingly important.

11.2.2 Live VM (Virtual Machine) Migration

Today VM (Virtual Machine) migration is common due to the ability to improve utilization of resources, load balancing of processing nodes, isolation of applications, tolerating the faults in virtual machines, to increase the portability of nodes and to rise the efficiency of the physical server. Likewise, live VM migration refers to moving these virtual machines around in a virtual environment, e.g., cloud data center. Strategies for live migration include accessing higher bandwidth for faster migration, configuration strategies, and the use of clusters to achieve these goals. The key thing about live migration is that this term implies that everything is being done without interruption to the running servers [2].

However, with all the enhanced techniques that ensure a smooth and flexible migration, the downtime of any VM during a live migration could still be in a range of few milliseconds to seconds. But many time-sensitive applications and services cannot afford this extended downtime, and their clocks must be perfectly synchronized to ensure no loss of events or information. In such a virtualized environment, clock synchronization with minute precision and error boundedness are one of the most complex and tedious tasks for system performance.

The best available solution for such precision time synchronization as warranted by live VM and DDBMS is PTP-based network though two other proposals also exist: one is "White Rabbit" and the other is DTP (Datacenter Time Protocol). The former has some practical deployment while DTP remains a theoretical contention. We will discuss both later in this chapter.

As the modern data centers are designed in PoD (Point of Delivery) based framework, it is possible to create cluster for precision time-sensitive applications and non-precision time-sensitive application. For the former, PTP clients with hardware stamping capabilities can be installed in DDBMS servers while non-precision time-sensitive application can still use NTP for synchronization. The TOR (Top of the Rack) switch that connects racks of servers, herein DDB (Distributed Database) servers can be a transparent switch providing PTP through either a boundary clock or grandmaster depending upon network setup. The WAN/DCI cluster of switches can also be transparent switch connected to a boundary clock or grandmaster and connects to remote data centers as shown in Fig. 11.2. In most cases, hypervisor do provide support for PTP and thus precision time accuracy can be obtained for live VM migration at hypervisor level. However, in this setting network asymmetry could add to time error budget, thus increasing end-to-end time error budget.

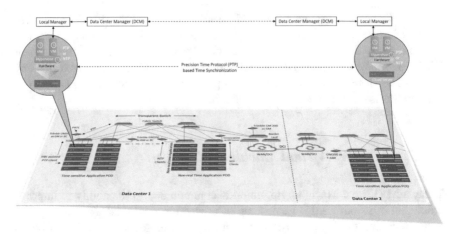

Fig. 11.2 Typical live VM migration in a PTP-based network

Henceforth, the boundary clock and PTP client installed in the server should account for the network asymmetry and provide asymmetry correction.

11.2.3   Time Synchronization for Hadoop Cluster

Enterprises who have embarked on *Big Data* journey know it well how vital it is to keep Hadoop clusters up and running. To meet expected SLA (Service Level Agreement)/OLA (Operational-Level Agreement) and to serve internal and external customers, it is of utmost importance to keep Hadoop clusters in good health. For that, one cannot ignore clock drift issues as discussed earlier. Ripple effect of clock offsets could have dire consequences, disrupting all services.

Hadoop is a master-slave architecture in which slave node sends regular heartbeat signals to master node regarding its presence and health. Thus, it is important that all the machines in the cluster are time synchronized and refer to the same time. Synchronization allows exchange of heartbeat between master and slave. If the synchronization is broken, it means no updates to master, about health.

Today, synchronization is commonly done using NTP server but there are many applications such as real-time big data analytics that depend on Hadoop framework requiring precision time synchronization. It is thus advisable that data center operators take an application-centric approach to deploy appropriate time synchronization for associated PoD. In this case, Hadoop cluster can be part of precision time-sensitive applications PoD.

11.3   Datacenter Time Protocol (DTP)

Recently, a number of scholarly articles [3, 4] posited the Datacenter Time Protocol (DTP) as a possible alternative to PTP to attain nanosecond-level accuracy without the issue of network asymmetry as it is applicable to packet-based timing protocol such as PTP. The postulations of DTP claims that the protocol can achieve 25 ns for directly connected nodes, about 150 ns for a data center with six hops and 200 ns for end-to-end network. The DTP aims to achieve nanosecond-level precision with scalability in a data center network, and without any network overhead and this is done by running a decentralized protocol in the physical layer (PHY) [3, 4]. DTP assumes that two peers are already synchronized in the PHY in order to transmit and receive bitstreams reliably and robustly. In this case, the physical layer of peer at the receive path (RX) recovers the clock from the physical medium signal generated by the transmit path (TX) of the sending peer's PHY as shown in Fig. 11.3.

Since the DTP is implemented at PHY level, doing so it eliminates delay error from network jitters and software stack. The DTP proposed a sublayer at the 10GbE PHY specifically to IEEE802.3ae 10 Gigabit/S Ethernet standard. Figure 11.4 shows DTP proposal for adding a sublayer to 10GbE PHY layer.

The control logic of DTP in a network port includes an algorithm [4] and a local counter. The DTP-enabled PHY as depicted in the figure above is exactly the same as PCS layer from the standard except that it has DTP Control, DTP TX, and DTP RX sublayers. Specifically, TX DTP inserts a protocol message on the transmit path for which DTP RX receives the message on receive path and forwards to DTP control through a synchronization FIFO. The processing of DTP message happens as such that the upper layer is oblivious of these operations. Lastly, when an ethernet frame is processed in the PCS sublayer it is left untouched by the DTP sublayer. DTP can also work with external protocol such as PTP, in that a PTP server will periodically broadcast DTP counter and UTC time to other servers. While DTP proposal for extremely high precision synchronization capabilities across the

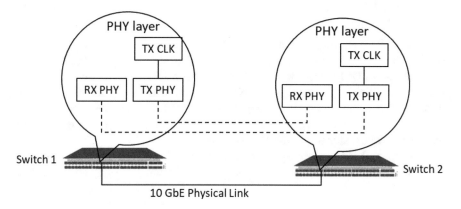

Fig. 11.3 Peer synchronization at PHY level for DTP-enabled device

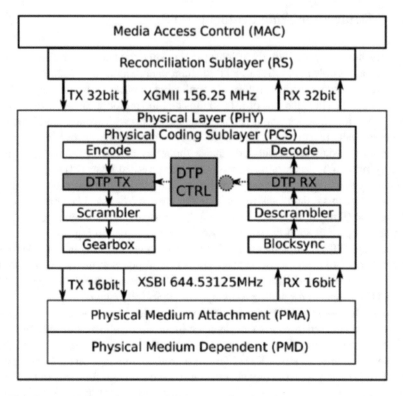

Fig. 11.4 Proposed DTP sublayer in IEEE802.3ae 10Gigabits/S Ethernet Standard [3, 4]

network is interesting, it remains a theoretical contention only and has not been implemented in the 10GbE PHY as of writing this book.

11.4   White Rabbit

White Rabbit (WR) [https://white-rabbit.web.cern.ch/] is a multi-collaborative project led by European Organization for Nuclear Research (known as CERN) with participation of GSI Helmholtz Centre for Heavy Ion Research and other partners from universities and industry. The WR protocol was originally developed to improve PTP for adapting it to the requirements of the particle accelerator industry and used by CERN for several years now. The WR protocol is gaining momentum and some vendors started to offer WR switches.

The protocol is based on SyncE defined by ITU-T G.8262 and PTP version 2 (IEEE1588-2008). It offers improvement over PTP through following objectives [5]:

- Synchronization and timestamping with sub-nanosecond accuracy.
- Clock frequency distribution with a precision better than 50 ps.
- Distribution through thousands of nodes and tens of kilometers over standard optical fiber networks. Specific network configuration is required for larger distances.
- Dependable and deterministic global time reference. Timing is not significantly affected by network traffic, weather conditions, or number of hops.

The default version of WR allows plug-and-play links up to 10 km with sub-nanosecond accuracy. Because of this, it is being adopted by many time-sensitive data center-centric applications, e.g., financial applications for which stringent clock synchronization is required. Recently, the WR protocol was adopted by Deutsche Börse stock exchange in Frankfurt to synchronize all their network. Several companies in the financial sector also implemented or in process of implementing WR to synchronize their network.

While PTP with hardware timestamping can achieve hundreds of nanoseconds to sub-microseconds precision, such accuracy may be impacted by network asymmetry issues. However, WR promises further improvement to PTP capabilities with sub-nanosecond accuracy over longer distance, typically 10 km and ability to serve more than 1 K nodes. To achieve sub-nanosecond accuracy a data center must deploy WR switches across the network. Figure 11.5 shows typical deployment of WR switches. Please note to achieve nanosecond-level accuracy at servers, PTP slave NIC must be installed. Some servers may require 1PPS input and as such can be achieved through implementation of timing module (e.g., Trimble RES720) with PTP slave NIC.

The WR switches work in a master slave configuration: at the highest level of hierarchy the WR master receives PTP, SyncE, 1PPS, and 10 MHz input from PTP grandmaster as shown in the figure above. Alternatively, the WR master can be configured as PTP grandmaster taking SyncE, 1PPS, and 10 MHz input from GNSSDO and/or cesium clock.

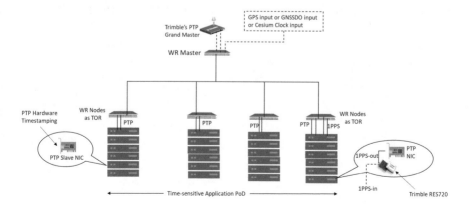

Fig. 11.5 Typical WR setup in a data center serving time-sensitive application PoD

WR switches allow users to build highly deterministic data center networks with layer 2 prioritization mechanism as defined by IEEE 802.1Q for sync traffic. The combination of deterministic latencies and a common notion of time to within subnanosecond allows WR to be a suitable technology to solve many problems in distributed real-time controls and data acquisition [6]. Currently, WR switches are available with 1GbE interface but CERN plans to introduce the specification for 10GbE WR prototype switches by December, 2020 with plan for 10GbE WR final release by March, 2024.

The layer 1 syntonization of WR switches depends on SyncE which guarantees frequency level synchronization of all nodes/switches in the network that supports SyncE. Such synchronization has no impact whatsoever on data traffic. We have discussed the functions and operations of SyncE in Chap. 6. In a SyncE setup, a downlink port of a switch disciplines the frequency of a downstream switch by connecting to its uplink port. The downstream switch uses the extracted clock to encode the data which it sends back to the upstream switch. This mechanism allows upstream switch to find a delayed copy of its encoding clock in the output of the CDR (clock and data recovery) circuit in its RX path, as depicted in Fig. 11.6.

The phase shift between these two clock signals is directly related to the link delay, and a measurement of this phase difference can therefore be incorporated into the PTP equation in order to achieve better precision. Additionally, a phase-shifting circuit can be included in the WR slave node to create a phase-compensated clock signal, i.e., a clock signal which is in phase with the master clock signal despite the delay introduced by the fiber link. The delay programmed into this phase shifter at any given time is of course taken into account when calculating the link delay [6]. The combination of layer 1 syntonization as in SyncE and phase tracking mechanism allows WR synchronization mechanism to operate independent of the data link layer, herein PTP (that operates at data link layer). Therefore, beyond the first PTP message exchange, there is in principle no need to continue exchanging PTP messages to keep the nodes synchronized. In fact, WR protocol implementation keeps the PTP messages going for robustness reasons, but at a much reduced rate making PTP exchanging negligible in terms of bandwidth consumption. This operation does not hinder determinism for the user frames, another key requirement for WR.

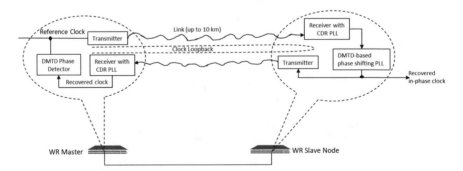

Fig. 11.6 The phase tracking mechanism in WR switches [6]

11.5   Application-Centric Sync Plane Design

NTP is predominantly used in data centers and some applications are tolerant to millisecond level time accuracy NTP provides. As discussed in Chap. 7, some hyperscale data centers are augmenting their NTP-based deployment by replacing Chrony daemon instead of ntpd. As evident by Facebook deployment, chrony daemon is relatively better choice than ntpd and significantly improves accuracy for NTP. While this may address the sync requirements for some applications and even allow Hadoop cluster to operate, it is not by no means suffice the need for highly precision time synchronization needed for many financial applications and DDBMS.

However, it is also cost-effective to keep NTP where possible but design the network such that applications and their node deployments are group together as per time tolerance. Modern data centers are designed in PoD (Point of Delivery) that is ideal for cluster-centric approach and thus application servers can be grouped together to an application-centric PoD. Doing so provides several benefits to data center as east to west traffic can be limited to where possible within a PoD or multiple PoDs and even isolated from rest of the data center traffic. Moreover, appropriate synchronization can also be applied to this application-centric PoDs. As such applications that are tolerant to millisecond level accuracy are grouped together and separated from those requiring nanoseconds or microseconds level accuracy.

As depicted in Fig. 11.7, time-sensitive and real-time applications are separated from non-time-sensitive applications each having their own PoD. This application-centric sync plane design is cost-effective and time accuracy is delivered where needed. Moreover, east west traffic is also localized. While such design can be effective for most application given that servers implement PTP slave NIC for hardware-based timestamping. This design can be further improved using SyncE-enabled whitebox as TOR (Top Of The Rack Switch) or WR switch nodes as TOR if needed. One drawback for WR nodes is that currently only 1GbE link speed is

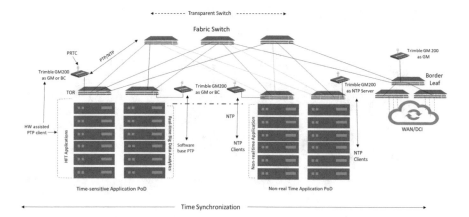

Fig. 11.7 Application-centric PoDs for data center sync plane design

supported. However, connecting TOR switch to grandmaster or boundary clock (specifically PRTC-B for T-GM and Class C for T-BC) will effectively improve time accuracy to possibly less than 1 μs. If WR deployment is done with GNSSDO where WR master acts as T-GM and WR nodes are connected to TOR instead of replacing it, accuracy may further improve to sub 10 ns or better.

11.5.1 Localized PRTC (Primary Reference Time Clock)

The network is prone to packet delay variations (PDV) which adversely affects high-precision time synchronization, thus minimizing such condition is imperative for time-sensitive applications. This is what makes application-centric sync plane design so important. With localized PRTC, a guaranteed UTC is available to time-sensitive applications irrespective of network condition and sub μs to tens of μs accuracy of time distribution is possible under the durance. Localized PRTC also offers multi-constellation services for failover so as to provide resilient timing. In case of a failure of one satellite, let's say GPS, many other GPS and GNSS (e.g., Galileo or GLONASS satellites) are available to provide a PRTC for UTC input as depicted in Fig. 11.8.

Moreover, where individual GNSS antenna cannot be installed, an antenna splitter can be used to create multiple feed from a single GNSS antenna and provided as input to multiple T-GMs (grandmaster clock) at different application cluster PoDs. Such design creates localized PRTC (Primary Reference Time Clock) for each application PoD for which frequency distribution to server can be done easily in addition to PTP.

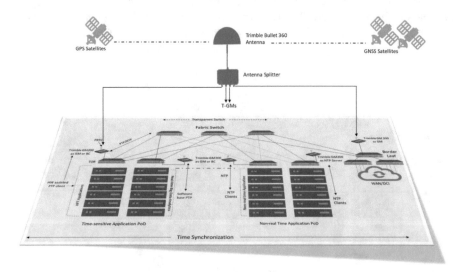

Fig. 11.8 Multiple T-GMs (grandmaster clock) supported by single antenna for GPS/GNSS time input

11.5.2   Sync Plane Overlay for Distributed High-Precision Synchrony

Due to network asymmetry and security issues, it is important to isolate sync traffic and prioritize it accordingly for distributed high-precision synchronization. This type of design for distributed high-precision synchrony can be easily performed given a mindset is created for application-centric sync cluster as discussed earlier. Today, L3-based fabric with spine/leaf physical architecture is common in data centers specifically to operators who requires scale out capabilities. For such network design as depicted in Fig. 11.9, a sync overlay can be easily created on top of IP underlay.

Many whitebox and/or Layer2/Layer3 switches support port buffering, advance QoS mapping and VXLAN/NVGRE protocols. In majority of spine/leaf architectures for scale out design eBGP is used with EVPN as control plane, as such VXLAN tunnel can be created from each TOR dynamically and on demand basis. MP-BGP EVPN control plane will create dynamic overlay tunnel and map the port of TOR that connects boundary clock directly to specific VNI. This way, communication between T-GM and T-BC can be done in a given overlay tunnel. Though the diagram above shows such tunnel within a data center and in between two application clusters, such design is very effective to create Time as a Service (TaaS) that can be extended to edge data centers or other regional data centers.

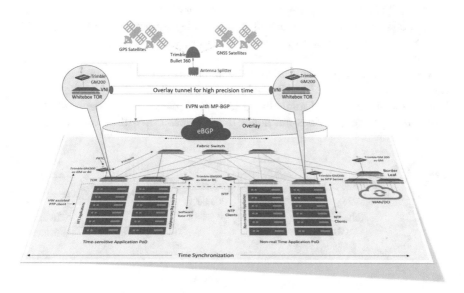

Fig. 11.9 Sync plane overlay for distributed high-precision synchrony

11.6   Synchrony for MSO (Multiple-System Operator) Infrastructure

From the early days, cable networks required good synchronization since the physical transmission medium is shared by cable modems and poor synchronization could cause interference and crosstalk. Secondly, moreover its existence with early days TDM network services required good synchronization. According to DOCSIS 1.0 specification dated 2001 that defines the "Data Over Cable Service Interface" and operations of CM (Cable Modem) and CMTS (Cable Modem termination system), time synchronization is inherent to physical layer functioning of RF interface. For example, from early days of DOCSIS 1.0 communications between CMTS and CM required time synchronization. Figure 11.10 shows typical cable network setup for IP communications. Essential problem is that a number of CMs share a single upstream channel (there are a number of upstream channels, and each channel can be treated independently, but even so, each upstream channel is shared by several modems).

In such setup, due to this shared medium/channels, there has to be an arbitrated mechanism by which CMTS provides an opportunity to a CM to transmit. Since that CMs and their controlling CMTS share a notion of time, this makes cooperation possible. The upstream is treated as contiguous sequence of mini-slots. A mini-slot is power-of-two multiple of 6.25 µs increments, i.e., 2, 4, 8, 16, 32, 64, and 128 times of 6.25 µs [7]. Table 11.2 depicts mini-slot to time-tick values as presented in DOCSIS 1.0 [7].

For the upstream transmission to occur proper synchronization is needed, thus the CMTS and the CM need to have an accurate idea of the correct time offset known as "ranging offset." Figure 11.11 shows how the time request is done and ranging offset occurs. The ranging offset is the delay correction applied by the CM to CMTS upstream frame time at the CM in order to synchronize the upstream transmission.

The CMTS transmits (on the downstream channel) a MAC management message, the MAP message or MAP PDU (Protocol Data Unit), that describes exactly how an upcoming series of mini-slots is to be used (please refer to figure above). If a multifunctioning CM does not honor this command, it will be discovered and

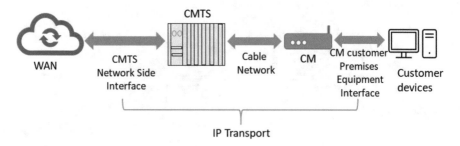

Fig. 11.10 Typical cable networks operated by MSO [7]

Table 11.2 Mini-slot to time-tick values as defined in DOCSIS 1.0 [7]

Parameter	Example Value
Time tick	6.25 µs
Bytes per mini-slot	16 (nominal, when using QPSK modulation)
Symbols/byte	4 (assuming QPSK)
Symbols/second	2,560,000
Mini-slots/second	40,000
Microseconds/mini-slot	25
Ticks/mini-slot	4

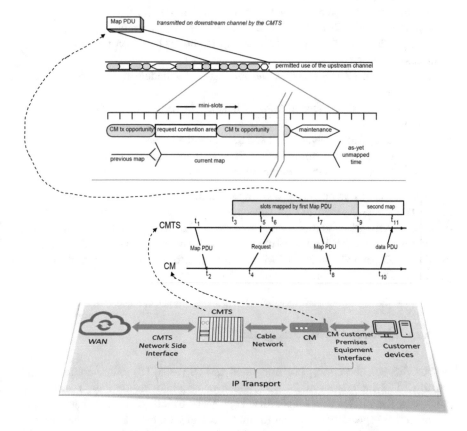

Fig. 11.11 CM and CMTS interaction granting upstream bandwidth and mini-slot allocation [7]

shutdown. A typical MAP might grant some mini-slots for the exclusive use of particular modems that have indicated in prior request frames that they have data ready to transmit requiring a number of mini-slots to transmit. The MAP PDU in the top part of the above figure shows how upstream bandwidth is divided into mini-slots and each mini-slot is numbered relative to a master reference maintained by CMTS. The bottom part of the figure above depicts typical communications between CM and CMTS once a CM is ready for data transmission. The basic operation of this communication is given as follows:

- *Step 1:* When a CM has a data PDU (Protocol Data Unit) to transmit, CMTS send map PDU at t1 with its effective starting time at t3. The MAP message includes a Request IE (Information Element) that will start at t5. The difference between t1 and t3 is needed to offset the downstream propagation delay allowing all CMs to receive the MAP, the processing time for CM, and the upstream propagation delay for first data from CM to arrive at CMTS.
- *Step 2:* At t2, CM receives the MAP, scan it for request opportunities. Additionally, to minimize collision a CM calculates t6 as a random offset.
- *Step 3:* At t4, the CM transmits the request for as many mini-slot is needed to accommodate the data PDU. The time t4 is chosen based on ranging offset as discussed earlier.
- *Step 4:* At t6, CMTS receives the request and schedule it for service in the next MAP.
- *Step 5:* At t7, CMTS transmit MAP (whose effective starting time is t9) with data grant for the CM.
- *Step 6:* At t8, CM receives MAP and scans for data grant.
- *Step 7:* At t10, CM transmits its data PDU so it arrives at the CMTS at t11. Time t10 is calculated from the ranging offset as in step 3.

Explanation of this operation was important to elaborate the case for synchrony in MSO networks. As evident from this operation that time synchronization is inherent to cable networks and it is not a choice but de-facto requirement to the functioning of cable networks. However, modern cable network requires synchronization consideration beyond CMTS to CM operation and for end-to-end network to support today's and future communications needs.

11.6.1   DOSIS Timing Interface (DTI)

In DOCSIS 2.0, the concept of integrated CMTS (I-CMTS) gains momentum in which CMTS included a 10.24 MHz clock or oscillator inside the chassis. This clock was the master timekeeper that kept every event synchronized with other events. It was especially important for microsecond level calculations necessary for DOCSIS transport: remember tick of 6.25 μs discussed earlier. Well, life was easier with I-CMTS but that got changed with DOCSIS 3.0 that provided the concept for distributed CMTS known as M-CMTS (Modular CMTS). This distributed architecture splits the components of a traditional CMTS or I-CMTS into an M-CMTS core that is connected via an interface to an edge QAM (EQAM). Additionally, DOCSIS 3.0 added enhancements to the prior DOCSIS standards such as channel bonding, support for IPv6, and support for IPTV. The channel bonding provided MSO with a flexible way to significantly increase downstream speeds to a minimum of 160 Mbps, and upstream throughput up to a minimum rate of 120 Mbps to customers.

With M-CMTS architecture as depicted in Fig. 11.12, it is possible to host CMTS core in a centralized location, say the headend and its functional components such as Edge Quadrature Amplitude Modulator (eQAM) and upstream receivers in remote hubsites. Essentially networking and MAC functions of CMTS remained with M-CMTS core and downstream PHY functions are incorporated in the eQAM. The main issue comes from 10.24 MHz clock that was part of original I-CMTS no longer a valid option for distributed architecture. Although it is possible to have three separate fee running clocks or extend a central clock input with separate cabling to three different components of M-CMTS, as such it is not a viable option given distance, cost, technical constraints, and environmental factors.

Therein comes the value of DTI (DOCSIS Timing Interface) which was defined by Cablelabs [9] with distributed architecture in mind and a means for distributing

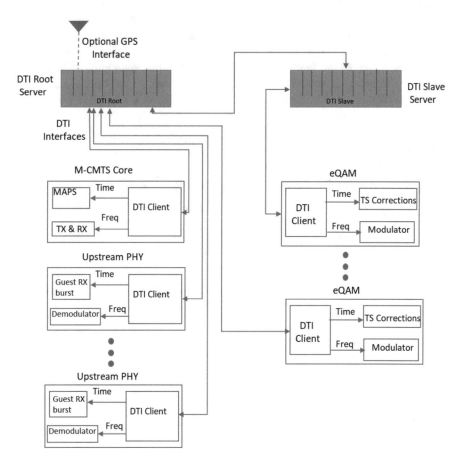

Fig. 11.12 DOCSIS Timing Interface (DTI) with M-CMTS core and redundant Upstream PHY and eQAM settings [8]

clocks among three components of M-CMTS. The DTI interfaces with redundant Upstream PHY and eQAM as presented in ITU-T J.211. The DTI protocol and components support accurate and robust transport of the server 10.24-MHz master clock and 32-bit DOCSIS timestamp to the client within a node or building [8]. The protocol is designed to minimize complexity of client clock cost and shared server costs along with support for all DOCSIS S-CDMA, TDMA, and future TDM services timing requirements in a distributed architecture.

The mechanism can achieve high-precision accuracy of <5 ns by using simple ping-pong layer-2 timing protocol over a single twisted-pair connection using common passive PHY components in both directions. The DTI components include DTI root server and DTI slave server and connectivity between DTI interfaces for CMTS components connectivity. The DTI root server provides a 10.24 MHz clock source while the DTI slave server works as redundant 10.24 MHz clock services in case of primary server failure. Figure 11.13 shows a DTI protocol communication between DTI root server and DTI client (M-CMTS functional components).

The first step in DTI operation is that DTI root server warmup and send DTI message with warmup state indication. At this point DTI root server will stay in the warmup until the oscillator is stable and a stable time of day setting is acquired and server transitions to the free-run state. It will stay in this state until locked to an external reference. As shown in figure above, client power up at timeslot M and wait for DTI server normal condition indication message. At M + 1 timeslot, DTI server

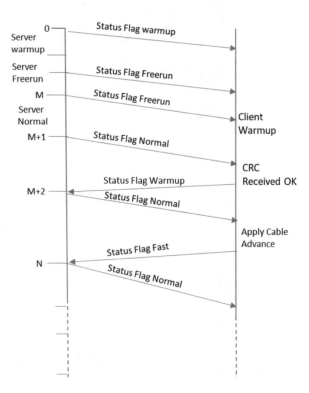

Fig. 11.13 A diagrammatical representation of the communications between DTI server and DTI clients [9]

indicates its normal condition that follows a series of status flag exchange between server and client from M + 2 timeslots onward. The figure above presented a typical communication with partial message exchanges as an illustration of the type of DTI communications between root server and DTI clients through which DTI client gets clock synchronized.

11.6.2   Remote DOCSIS Timing Interface (R-DTI)

We learned about M-CMTS as first step towards distributed architecture and implementation of DOCSIS 3.0 specification supporting the distributed architecture of M-CMTS and multiservice cable networks by MSO. This distributed architecture concept got another boost through DOCSIS 3.1 that defined high-speed communication 10 Gigabits/S for downstream and 1 Gigabits/S for upstream along with advanced OAM functions, improved error correction and support for modern network architecture, e.g., SDN and SD-WAN. Along the way came the evolution of CMTS platform to CCAP (Converged Cable Access Platform). It provided option for manufacturers to achieve increased Edge QAM and CMTS densities that MSOs require by leveraging existing technologies such as DOCSIS and also newer technologies such as Ethernet optics and EPON (Ethernet Passive Optical Network). In summary, CCAP unifies the CMTS, Switching, Routing, and eQAM functions at the headend, so that all data, video, voice functions can be handled over IP before conversion to RF or Optical signals [10].

Soon after DOCSIS 3.1 release, Cablelabs also introduced R-PHY (Remote PHY) in 2015 to further enhance distributed architecture of cable network. The R-PHY pushes the physical RF layer (PHY) to the edge of the access network. This type of deployment essentially split CCAP into two parts: MAC and PHY layers. The CCAP MAC forms the CCAP Core that includes CMTS Core and eQAM Core. The PHY layer is pushed to the edge of the network and thus known as Remote-PHY or R-PHY as shown in Fig. 11.14. The device that implements R-PHY is known as RPD (Remote PHY Device). Such deployment allowed MSO networks to scale out delivering on demand services that otherwise was difficult. Moreover, this design introduced Ethernet and Ethernet Passive Optical Network (EPON), specifically 10GbE to 100GbE for Ethernet link and 10GbE EPON to NGPON2 for PON (Passive Optical Network).

However, with this new design came the challenge for synchronization as it relates to precision time distribution along the Distributed Access Network (DAC) path. To solve this issue, Cablelabs introduced Remote DOCSIS timing interface (R-DTI) specification in 2014 [11] supporting PTP (defined by IEEE1588) between CCAP core and R-PHY and DTP (DOCSIS Timing Protocol) between R-PHY and CM as shown in the figure above. Whereas Ethernet based networks concern, PTP allows both phase and frequency information to be transferred between nodes across the packet network with switches or routers, thus making it ideal for R-DTI. However, the usage of PTP across asymmetrical links, such as EPON, is problematic since

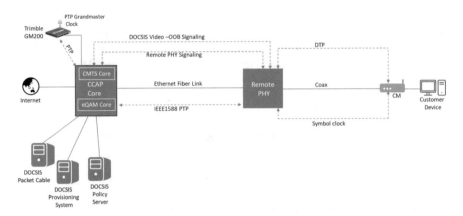

Fig. 11.14 A diagrammatical representation of CCAP and R-PHY setup with timing interfaces

IEEE1 588 relies on the basic premise that the link delays are symmetrical. Hence, to address this issue for EPON links, use of IEEE 802.1AS standard is recommended [11], specifically the method for transferring time and frequency information across an EPON link. It is to be noted that IEEE 802.1AS utilizes PTP as core timing protocol but it specifies IEEE 802.3 Multi-Point Control Protocol (MPCP) for use with EPON that is not part of PTP profile.

For typical EPON deployment, OLT (Optical Line Terminal) and ONU (Optical Network Unit) form the access path over a single fiber that is split to many fiber strands using optical splitter for data transport. The primary function of the OLT is to convert, frame, and transmit signals for the PON network and to coordinate the optical network terminals multiplexing for the shared upstream transmission. The ONU on the other hand converts optical signals transmitted via fibers to electrical signals which then sent to subscribers. In summary, while OLT connects to core network for data transport and allows single fiber to be shared, ONU extends the reach of fiber and data transport to last mile. For the simplicity of explanation, the discussion below considers OLT implemented at CCAP Core while ONU implemented at the R-PHY devices (RPDs).

With this assumption, R-DTI implementation in an EPON environment can be achieved through a combination of PTP and IEEE802.1AS MPCP. For this, PTP is terminated at the CCAP Core and IEEE 802.1AS MPCP is used to transfer timing from the CCAP Core to the ONU element in the RPD, as shown in Fig. 11.15.

The DTP between RPD and CM allows the timing and frequency system of DOCSIS to be interfaced to external timing protocols such as PTP with high accuracy. According to R-DTI specification [11], RPD must support DTP for the applications that require precision synchronization, e.g., a 4G/LTE eNB that can be connected to the CM. Whereas a mobile fronthaul connectivity is concern, end-to-end time error budget does not change and remains at 1.5 μs from PTP interface at T-GM connected to CCAP to eNB connected to CM.

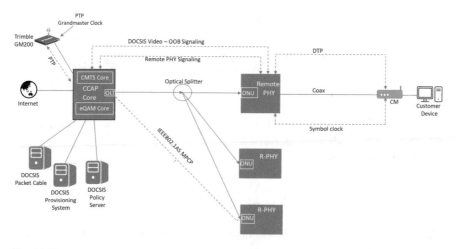

Fig. 11.15   R-DTI implementation using IEEE802.1AS MPCP for EPON deployment

11.6.3   The DTP (DOCSIS Timing Protocol)

The DOCSIS Timing Protocol (DTP) was introduced as part of DOCSIS 3.1 to provide enhanced timing support in a DOCSIS-based cable network. As defined in DOCSIS 3.1, "DTP is a set of techniques coupled with extensions to the DOCSIS signaling messages which allow the timing and frequency system of DOCSIS to be interfaced to external timing protocols with high accuracy" [12]. The primary objective of DTP is to provide precise frequency and time to an external system that is connected to the network port of a DOCSIS CM.

In a typical deployment, a CMTS or CCAP may have PTP or DTI input for frequency and time source or it may integrate the function of T-GM. In either case, DTP allows these timing sources to be accurately replicated at the egress port of the CM. This is accomplished by combining a set of native DOCSIS protocols such as downstream frequency recovery and time synchronization with DTP signaling and DTP math to allow compensation for asymmetry in network and processing delays. The protocol relies on the DOCSIS 3.1 "Extended Timestamp" and defines five categories (we will discuss this in DTP signaling section) of system timing accuracy with time synchronization error between two CMs in range from 100 to 3000 ns [12]. Figure 11.16 depicts a typical setting of DTP where various timing source at CMTS is accurately replicated and transported to CM for accurate timing output.

In defining the DTP architecture which includes clock elements of CMTS core and CM, DOCSIS 3.1 also specifies that the entire architecture construe as boundary clock in a native PTP environment. This assertion innately suggests that time error budget between CMTS and CM clocking elements should be less than 100 ns so to comply with ITU-T G.8273.2 max TE for boundary clock class A. This applicability of boundary clock as defined in ITU-T G.8273.2 should be applicable to

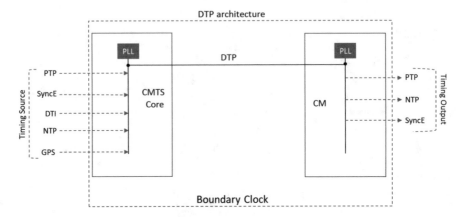

Fig. 11.16 The DTP architecture showing implementation of DTP between CMTS Core and CM

telecom and other environment where precision time synchronization is required, and end-to-end time error budget is stringent.

11.6.3.1 DTP Signaling

The goal of DTP is to generate a time adjustment (t-ad) that can be added to the native timestamp of the CM to create a timestamp that matches the CMTS time-stamp in real time [12]. The DTP time calculations can be provided by either CMTS or the CM as needed, in that case the entity performing the DTP calculation is known as DTP master while other partner entity is DTP slave. For timing transaction, DTP master initiate the communications and DTP slave respond to the message. A series of DTP messages are being exchanged between DTP master and DTP slave. The figure below shows CMTS as DTP master and the CM as DTP slave. The DTP master initiated first time transaction message "DTP-REQ" that includes clock id, CMTS, and other parameters. The DTP-REQ message is responded by DTP slave with DTP-RSP that includes CM parameters and TRO (True Ranging Offset). In response of DTP-RSP, DTP master sends DTP-info that includes time adjustment (t-ad). The DTP Slave respond this with DTP-ACK for acknowledgement.

The timing transaction presented in Fig. 11.17 which includes the transfer of the external timing protocol from CMTS to CM may introduce timing errors in the form of latency, jitter, and skew. Although the latency is managed and compensated by DTP protocol, there can still be a latency error between two systems. The combined system timing errors due to latency variation, jitter, and skew are described in Table 11.3.

However, it is to be noted that jitter and skew budget depends upon the desired accuracy of the DTP system. Table 11.4 shows five categories of DTP system discussed earlier and their corresponding error budget.

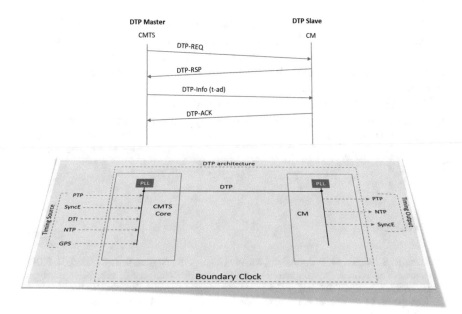

Fig. 11.17 The DTP message exchange for time synchronization

Table 11.3 DTP system parameters for jitter and skew [12]

Name	Description
T-cmts-error	This is the variance in delay introduced by CMTS and is measured from the clocking ingress port to the CMTS DOCSIS egress port
T-cm-error	This is measure for CM similar to the above: From DOCSIS ingress port to timing output port
T-docsis-error	Combined timing error. T-docsis-error = T-cmts-error + T-cm-error
T-source-skew	The max allowable difference in arrival time of a reference timing source ports of two CMTSs that exist within the same timing system
T-hfc-error	The latency error introduced by the modeling of the HFC plant
T-cm-cm-skew	T-cm-cm-skew = 2 * T-docsis-error + T-source-skew + 2 * T-hfc-error

Table 11.4 Five categories of DTP system and the corresponding time error budget [12]

Parameter	Level I	Level II	Level III	Level IV	Level V
T-cmts-error	±20 ns	±40 ns	±150 ns	±200 ns	±500 ns
T-cm-error	±20 ns	±40 ns	±200 ns	±300 ns	±500 ns
T-docsis-error	±40 ns	±80 ns	±350 ns	±500 ns	±1000 ns
T-source-skew	5 ns	10 ns	100 ns	200 ns	500 ns
T-hfc-error	±7.5 ns	±15 ns	±50 ns	±150 ns	±250 ns
T-cm-cm-skew	100 ns	200 ns	900 ns	1500 ns	3000 ns

The Level I system is considered a DTP system with built-in PRTC or GPS input, whereas Level II system is driven by relaxed positioning requirements. The Level III and Level IV DTP system are driven by LTE-A macro and small cell with higher TE budget. The Level V DTP system requirement is driven by typical DOCSIS deployment.

This DTP system definition clearly indicates that selection of DTP for particular implementation depends upon application requirements. For example, 5 g fronthaul may require far more stringent DTP system TE requirement than those specified in the table. However, Level I DTP system TE may be acceptable for certain 5G deployment.

References

1. Khan, R. S., Malik, M. S. A., Ashraf, M. W., Ullah, A. S., Asghar, I., & Razzzaq, N. (2019). A review on synchronization and concurrency control techniques of distributed databases. *IJCSNS International Journal of Computer Science and Network Security, 19*(2).
2. Techopedia. (2020). *What is the difference between vMotion, VM migration and live migration?* Techopedia.com.
3. Lee, S. K., Wang, H., Shrivastav, V., & Weatherspoon, H. (2016). *Globally synchronized time via datacenter networks.* Cornell University.
4. Lee, S. K., Wang, H., & Weatherspoon, H. (2019). Globally synchronized time via datacenter networks. *IEEE/ACM Transactions on Networking, 27*(4).
5. DOT. (2019). *Seven solutions response to department of transportation RFI for demonstration of backup and complementary positioning, navigation, and timing (PNT) capabilities of global positioning system (GPS).* Agency/Docket Number: DOT-OST-2019-0051. US Department of Transportation.
6. Serrano, J., Cattin, M., Gousiou, E., van der Bij, E., Włostowski, T., Daniluk, G., et al. (2013). *The white rabbit project.* IBIC.
7. DOCSIS 1.0. (2001). *Data-over-cable service interface specification 1.0. Radio frequency interface specification.* SP-RFI-C01–011119. Cable Television Laboratories, Inc.
8. J.211. (2006). *Timing interface for cable modem termination systems.* ITU-T J.211. International Telecommunication Union.
9. DTI. (2008). *Data-over-cable service interface specifications modular headend architecture.* DOCSIS timing interface specification: CM-SP-DTI-I05-081209. Cable Television Laboratories, Inc.
10. Sundaresan, K. (2015). Evolution of CMTS/CCAP architectures. In *2015 spring technical forum proceedings.* CableLabs.
11. DCA - MHAv2. (2014). *Data-over-cable service interface specifications: DCA - MHAv2.* Remote DOCSIS timing interface: CM-SP-R-DTI-I07–180509. Cable Television Laboratories, Inc.
12. DOCSIS 3.1. (2013–2020). *Data-over-cable service interface specifications: DOCSIS® 3.1.* MAC and upper layer protocols interface specification: CM-SP-MULPIv3.1-I21-201020. Cable Television Laboratories, Inc.

Chapter 12
Synchronization for Industrial Networks

12.1 Introduction

With increased push for industry 4.0 that innately allows digitization of industrial control system, integration IIOT (Industrial IOT), cyber physical systems, smart manufacturing, big data and holistic digital management of industrial systems, industrial networks are more important now than ever before. However, such integration of disparate ecosystem to a connected framework demands more than just interconnections of devices. It requires communications protocols that can work with otherwise incompatible industrial control systems, and provides guarantee of transmission and timeliness of data transport. To safely and efficiently operate industrial systems, determinism and real-time control are two key imperatives that industrial network must offer. However, there is no single technology that dominates industrial network. Some technologies have better market share than others in industrial networks yet users have not settled on a single communication technology for industrial communications. According to recent market reports, Ethernet technologies are increasingly penetrating in the industrial networking marketplace taking away market share from fieldbus technologies. These Industrial Ethernet (IE) technologies are also having varied transport mechanism and more importantly, each has their own value-add for determinism and real-time control in a given industrial network. It is however important to note that industrial networks are time critical unlike standard networks that are deployed elsewhere. Industrial network needs high-precision time synchronization and inaccuracy may render in catastrophic failure in certain industrial environment. In this chapter, industrial networks are explored under two distinct categories: Industrial Ethernet and Fieldbus technologies. Each of these categories offer a set of technologies that are equally important for some industrial use cases and each offers distinct value-adds for specific problem they solve while connecting industrial control systems. Later, synchronization aspects of these technologies are examined with specifics on deployment scenarios.

12.2   Fundamentals of Industrial Networks

As a backbone of automation system interconnect, industrial networks (sometimes refers to as industrial control networks) provide a powerful means of data exchange, data controllability, and flexibility to connect various industrial systems and devices. Unlike traditional communications network, it is a highly deterministic network with ability to provide real-time control. The word deterministic means the underlying network provides latency guarantee and ultra-low packet loss. On the other hand, real-time control implies that network provides a means of coupling with tighter synchronization. In the context of industrial automation and drive technology, real time means that connected industrial systems and devices are able to safely and reliably reach cycle times in the range of less than ten milliseconds down to microseconds.

In recent years, IETF working group has undertaken initiatives to define the deterministic network and its operation through implied guidance for layer 2 and layer 3 protocols operations. A series of RFCs (e.g., RFC 8557 and RFC 8578) are released by IETF DetNet (Deterministic Networking) working group that will shape the future of industrial networks. The DetNet initiative however does not define the synchronization for the network, instead it recommends the use of IEEE802.1AS standards. For most control applications, information transmission or receipt is not time sensitive. However, for faster applications such as motion control, the very nature of the control loop requires a predictable amount of time between the receipt of new data for it to work properly. This control scheme requires a deterministic network: one for which the exact time to expect the message is known and predictable. For time-sensitive variables in any control loop, such as motion control, a message arriving late can be very bad news [1]. Hence, an industrial network must guarantee low latency and ultra-low packet loss with tighter coupling for time-sensitive operations of industrial systems. Moreover, today's industrial systems are required to provide enhanced production monitoring and quality control and at the same time maintaining the operation costs as low as possible. The advancement of information and communication technologies has enabled the industrial systems to match up with these needs. Such advancements led to significant reduction of manual labors replaced with faster and more reliable automated machines in the most of industry operations. It also provides both the factories and the manufacturing plants with necessary monitoring which they both sought for better supervisory and quality control. Unification of these automated machines and devices into the factories and other industrial environments need an efficient method to connect them together, to communicate with each other, and to transfer the various supervisory data to the monitors. These imperatives steered the development and deployment of a deterministic network such as industrial network for the industrial sectors [2].

To enable these industrial control functions, industrial networks provide inter-connect for three distinct components: Programmable Logic Controllers (PLC), Supervisory Control and Data Acquisition (SCADA), and Distributed Control Systems (DCS). The PLC is nothing, but a digital computer specifically designed for hazardous industrial environment. It continuously monitors the state of input devices and makes decisions based upon a custom program to control the state of output devices and then present the output to HMI (Human Machine Interfaces). On the other hand, SCADA is a monitoring and control software that typical manages PLC and record data for multiple segments including remote locations. It uses com-puters, networked data communications, and graphical user interfaces for high-level process supervisory management by relying on PLCs and discrete PID (Proportional, Integral, and Derivative) controllers to interface with the process plant or machin-ery. While a PLC is general purpose controller, PID is a special purpose controller typically used for control loop feedback to ensure the desired output for some kind of sensors, e.g., temperature sensor. The DCS is similar to PLC function in the sense that both are used to control from one or a few production processes. Historically, PLC was unable to coordinate the control of entire plant and that is the reason for the development of DCS. Thus, by definition DCS supervises and coordinates many systems of sensors, controllers, and associated computers throughout the plant.

In an industrial network, these functional elements are organized into a tiered topology in three basic levels: informational, control, and device level. Figure 12.1 depicts typical tiered architecture of industrial networks. There is unique

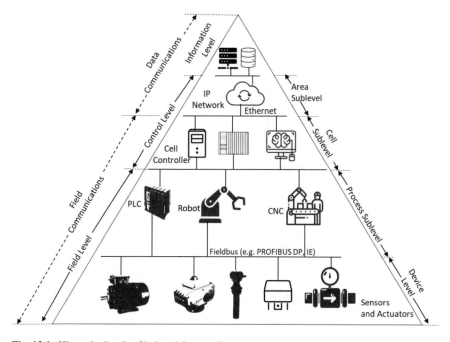

Fig. 12.1 Hierarchy levels of industrial network

requirement at each level and thus it affects which network to be used for that level. The field level has two parts device level and process sublevel for which device level consists of sensors and actuators while process sublevel includes PLC, CNC (Computer Numerical Control) machines and robotic arms, etc. The control level includes two distinct network elements: cell sublevel and area sublevel. The former includes SCADA and HMI for supervisory control, monitoring, and other control functions, and the latter includes communications via IP Network with Ethernet for underlying connectivity. Top of the pyramid is information level which includes WAN and systems to deal with large amount of data analysis and management, e.g., ERP system.

12.2.1   Communications Protocols

Depending upon the hierarchy level of industrial network, a series of often competing communications protocols are used for device and system connectivity. According to several market research reports [3, 4], Industrial Ethernet is projected to represent more than 64% and fieldbus represents 30% of all new nodes deployed in industrial networks.

Table 12.1 shows a various technologies group under Industrial Ethernet (IE) and fieldbus technologies.

It is to be noted that other than Ethernet TCP/IP, all protocols and technologies listed under "Ethernet" in the table above are using protocol level modification with

Table 12.1 Various industrial networks technologies representing three distinct group of protocols: ethernet, fieldbus, and wireless [3]

Group name	Communication technology	Market share (%)
Industrial Ethernet	EtherNet/IP	17
	PROFINET	17
	EtherCAT	7
	Modbus TCP	5
	POWERLINK	4
	CC link IE field	2
	Ethernet TCP/IP	12
Fieldbus	PROFIBUS DP	8
	Modbus-RTU	5
	CC-link	4
	CANopen	3
	DeviceNet	3
	Other fieldbus	7
Wireless	WLAN	3
	Bluetooth	1
	Other wireless	2

underlying ethernet medium for transport. These group of ethernet based protocols are collectively known as Industrial Ethernet, and not similar to standard Ethernet used elsewhere in the communications networks. Ethernet TCP/IP on the other hand represents standard Ethernet based IP network. For the simplicity of discussion, we will explore few leading Industrial Ethernet and Fieldbus protocols that will serve as the basis for better understanding of synchronization requirements in industrial networks.

12.2.2   *Industrial Ethernet (IE)*

Industrial environment such as factory floor has different environmental and communications needs than standard communications environment in enterprise networking. These requirements include physical security concerns, harsh environmental conditions, noise vibration, operational concerns, deterministic network needs, and real-time control. As such network including physical medium and communications protocols must maintain operational continuity without any performance degradation. With increased digitization in factories and other industrial environments, industrial automation systems are increasingly adopting ethernet as a choice of communications medium for the interconnectivity. The key operational aspects of IE are better determinism and real-time control. In order to achieve these, IE uses specialized protocols in conjunction with Ethernet. The more popular industrial Ethernet protocols are EtherNet/IP, PROFINET, EtherCAT, Modbus TCP, and Powerlink.

12.2.2.1   EtherNet/IP

The EtherNet/IP is defined by ODVA for which IP stands for industrial protocol as opposed to internet protocol. It is an application layer managed by ODVA that sits on top of the lower network layers. This allows the use of standard networking

Fig. 12.2 EtherNet/IP architecture

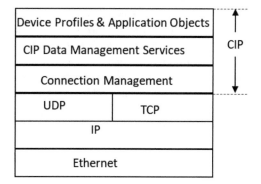

hardware. Figure 12.2 depicts EtherNet/IP architecture. It uses CIP™ (Common Industrial Protocol) encapsulated with TCP/IP header and packets are transported as part of TCP/IP datagram.

All devices in an EtherNet/IP network present their data to the network as a series of data values called "attributes" grouped with other similar data values into sets of attributes called "Objects." These objects are defined by CIP protocol and are of two kinds: required object and application object. The required objects are identity, TCP, Router, Connection, and EtherNet/IP objects. All EtherNet/IP devices must have these objects. Application objects on the other hand are organization of data that is specific to a device. Let's take an example of flow meter, by definition of its operation we understand that a flow meter should possibly have a flow object, and since it may collect the temperature data it should have temperature object. Similarly, it should also have input and out objects, etc.

For EtherNet/IP transport messages are of two kinds: explicit and implicit. Explicit messages are sent by the client (e.g., a controller or HMI) at any time, and the server (a field device, such as a servo drive) can respond when it is available. Therefore, explicit messaging is used only for information that is not time-critical, such as diagnostic or configuration data.

In contrast, implicit messages are used for real-time control. When time-critical information needed to be exchanged, EtherNet/IP device uses implicit messaging. In this operation, a control device (such as a PLC) establishes a connection referred to as a "CIP connection" with an adapter device (such as an actuator) and the information to be exchanged is identified when the connection is established.

12.2.2.2 PROFINET

PROFINET is an open Industrial Ethernet standard and a communication protocol that exchanges data between automation controller and devices. PROFINET uses three communications channel TCP/IP, PROFINET RT (Real Time), and Isochronous Real-Time (IRT). Like EtherNet/IP, PROFINET also uses TCP/IP (including UDP/IP), but only used for certain non-critical tasks such as configuration, parameterization, and diagnostics. However, PROFINET provides determinism for real-time control through PROFINET RT and IRT.

The PROFINET RT handles time-critical data exchange. It bypasses TCP/IP layers allowing the frames to arrive directly to PROFINET application. This helps eliminate delays induced by TCP/IP layer processing and speed up the communications, thus improving determinism significantly. PROFINET IRT is a step beyond PROFINET RT. It eliminates variable data delays by enhancing the rules employed to switch Ethernet traffic, and by creating special rules for PROFINET traffic. The protocol fulfills all synchronization requirements and allows deterministic communication. Specifically, PROFINET IRT enables applications having a jitter of less than 1 µs [5].

12.2.2.3 EtherCAT

EtherCAT is open IE standard for real-time control. It gets rid of the CSMA/CD mechanism that is inherent to Ethernet and replaces it with a new "telegram" message packet that can be updated on the fly. Networked devices are connected in a ring or a daisy chain format that emulates a ring. As data is passed around the ring, each node reads the data addressed to it and writes its data back to the frame all while the frame is moving downstream [6]. The EtherCAT telegrams are transported directly by the Ethernet frame with an ether type field identifier (Ox88A4). To ensure multi-hop communications, TCP/IP connections can optionally be tunneled through a mailbox channel without impacting real-time data transfer. This allows the network to offer a determinism of <30 μs with up to 1000 nodes [2].

12.2.2.4 Modbus TCP

The Modbus TCP is simply a TCP/IP encapsulation of Modbus frame which is an open serial protocol derived from the master/slave architecture originally developed by Modicon (now Schneider Electric). With a TCP interface that runs on Ethernet, Modbus takes advantage of popular ethernet based IE infrastructure for transport of Modbus RTU protocols.

The Modbus messaging structure defines the rules for organizing and interpreting the data independent of the data transmission medium. All Modbus commands and user data are encapsulated into the datagram of a TCP/IP without the Modbus checksum and transport relies on standard Ethernet TCP/IP link layer checksum methods for data integrity. Figure 12.3 depicts typical Modbus TCP Application Data Unit (ADU) encapsulation.

Please note, "Address" field from general Modbus frame is supplanted by "Unit ID" and becomes part of MBAP (Modbus Application Protocol) header in Modbus

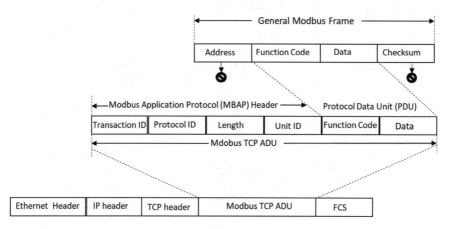

Fig. 12.3 TCP encapsulated Modbus TCP ADU

TCP ADU. Modbus TCP relies on underlying IP network for proper route to destination slave instead of using "slave id" as in standard Modbus operation.

12.2.2.5 POWERLINK

The POWERLINK is another deterministic technology for IE that was developed in order to be able to transmit data in the microsecond range. Originally developed by B&R, POWERLINK is an open standard Real Time Ethernet (RTE) protocol defined by Open user group EPSG. The protocol is mainly used for transmission of process data in automation technology. It enhances ethernet data link layer with mechanism to allow deterministic data transmission in which dedicated timeslots are given to isochronous and asynchronous data. Moreover, it replaces CSMA/CD with mechanism that only allows one networked device gain access to the network media at a time. It uses a mechanism known as "Slot Communication Network Management (SCNM)" to guarantee that isochronous and asynchronous data transmission will never interfere with each other. Figure 12.4 illustrates a layered architecture for POWERLINK.

As shown in the diagram above, the POWERLINK protocol uses Ethernet PHY and MAC layer while adding its own data link layer. This allows the POWERLINK

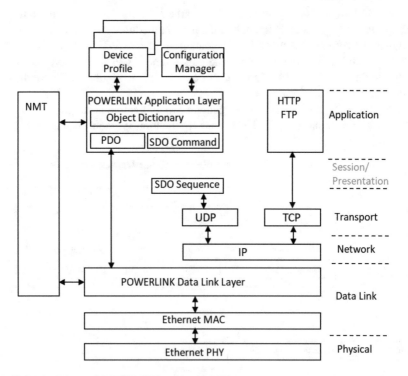

Fig. 12.4 The Ethernet POWERLINK architecture [7]

network to operate in two modes: POWERLINK mode and basic Ethernet mode. The POWERLINK mode allows for a Master Node (MN) to manage the network allowing only one node at a time to grant access to the network. It also specifies a set of rules for RTE (Real Time Ethernet) operation. In POWERLINK Mode, all message follows POWERLINK specific messages.

The operation of basic Ethernet mode is same as standard Ethernet operation (e.g., CSMA/CD) and mainly used for asynchronous traffic. The basic Ethernet mode uses UDP/IP or TCP/IP for transport.

The application layer comprises a concept to configure and communicate real-time data as well as the mechanisms for synchronization between devices. Each node in POWERLINK network exposes its interface to the "outside world" (i.e., the network) via a so-called object dictionary. The participants of the POWERLINK network can access the object dictionary of other nodes either asynchronously (Service Data Objects, SDOs) or synchronously (Process Data Objects, PDOs) [7, 8]. The Network Management (NMT) state machines defined the communications functions for both Master Node (MN) and Slave node known as Controller Node (CN).

12.2.3 Fieldbus

In early days, control systems in the factories and plants were analog in nature. These devices used direct connection between the controller and actuator or transducer through 4–20 mA control signal. As control system became complex and networking technologies evolved, fieldbuses were developed to interconnect the digital systems and devices. This led to a myriad of fieldbus protocols (Foundation Fieldbus H1, ControlNet, PROFIBUS, CAN, etc.) being developed over the last couple of decades. As a digital interconnect technology, fieldbuses are considered LAN technology with a single network cable replacing dozens and even hundreds of analog cables for the connectivity of sensors, actuators, and other devices to controllers. The fieldbus protocols allowed operators to easily monitor, control, troubleshoot, diagnose, and manage all devices from a central location. While these fieldbuses reduced the wiring and improved the reliability and flexibility of the system, another issue was created: multiple proprietary systems, with incompatibility and a lack of interoperability between the various components. Devices made to work with one fieldbus and protocol could not work with another [2]. A series of fieldbus technologies and protocols were developed to address the field level communications issues and create industry standards for communications at fieldbus level. However, no single standard emerged as the main conduit of field level communications. Table 12.1 depicts a list of major fieldbus technologies some of which will be briefly discussed here to provide a basic understanding.

12.2.3.1 PROFIBUS DP

The PROFIBUS DP (Profibus-DP) is a variation of PROFIBUS technology which is a standard for fieldbus communication in automation technology and was first promoted in 1989 by BMBF (German department of education and research) and then used by Siemens. PROFIBUS is not the same as PROFINET, it is a messaging format specifically designed for high-speed serial I/O in factory and building automation applications. Recognized as the fastest fieldbus in operation today, PROFIBUS is an open standard based on RS485 and the European EN50170 Electrical Specification. The DP suffix refers to "Decentralized Periphery," which is used to describe distributed I/O devices connected via a fast serial data link with a central controller. Normally, a PLC has its input/output (I/O) channels located centrally. This causes problem and cost associated with cabling throughout the plant to connect with sensors/actuators or I/O slaves. PROFIBUS DP allows decentralization of I/O channels bring it closer to sensors/actuators. This way, cabling can be minimized to the field area only and a network bus from IM (Interface Module) at I/O channels can extend the reach of PLC to I/O channels over a RS485 cable with PROFIBUS DP for communications to complete the connectivity.

12.2.3.2 Modbus RTU

The Modbus protocol was introduced in the 1970s by the American Modicon Company for PLC communications. It is an open standard real-time communication protocol that is widely used in controllers and measuring instruments. Today, it has becoming an international standard in the industrial automation field. The protocol supports traditional RS232, RS422, and RS485 interfaces as well as the Ethernet interfaces. It has two transmission modes: ASCII and remote terminal unit (RTU). In ASCII mode, the message is expressed by ASCII code and uses a longitudinal redundancy error check. In RTU mode, the message is expressed in binary codec decimal format and uses the cyclic redundant checksum (CRC) error check [2].

The Modbus RTU is widely used within Building Management Systems (BMS) and Industrial Automation Systems (IAS). The messages of Modbus RTU are simple 16-bit CRC (Cyclic-Redundant Checksum). This simplicity of the messaging structure ensures reliability.

In a Modbus RTU setup, the master initiates a request by using the function code, while the slave sends a response in reaction to that request. A successful response echoes the function code sent to it by the master. An exception response replies with the same function code but with the msb (most significant bit) set high to indicate something went wrong. Figure 12.5 depicts Modbus RTU communications between master and slave.

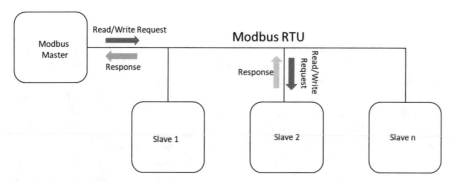

Fig. 12.5 The Modbus RTU network

12.2.3.3   CC-Link

The CC-Link (Control & Communication Link) is an open-technology fieldbus network originally developed by Mitsubishi Electric in 1996. It is widely used in Asia, but a large number of CC-link networks are also deployed in Europe and North America. The specification for this family of protocols is developed and maintained by CLPA (CC-Link Partner Association). The CC-link provides high-speed deterministic communication link to a wide range of multi-vendor automation devices over a single cable. This family of fieldbus network is suitable for machine, cell or process control in various industries ranging from semiconductor to food and beverage, automotive to pharmaceuticals, and material handling to building automation. CC-link family of fieldbus network technologies comprise of four different network types that together provide high-speed, deterministic communication, and integration of floor devices from controller to sensor level. These networks are:

1. *CC-Link*: This is a field device network that processes both control and data information at a speed of 10 Mbps to provide efficient integrated factory and process automation.
2. *CC-Link IE*: The CC-Link IE provides up to 1GbE transfer rate and uses deterministic IE for large volume of data transfer.
3. *CC-Link Safety*: This network is useful for safety applications that requires compliance with IEC 61508SIL3, EN 954-1 category 4 and EN ISO 13849-1 performance level. This network is compatible with CC-link network which allows the use of existing install base. The CC-Link Safety network has fail-safe function that will bring the machinery to a safe condition if a fault is detected.
4. *CC-Link LT*: This is a bit-oriented network designed for implementation in sensors and actuators, as well as for widespread I/O applications. The CC-Link LT can work as standalone or it can be connected to a CC-Link network.

12.2.3.4   CANOpen

Originally developed in 1983 for in-vehicle network, the CAN (Controller Area Network) or CAN bus still dominates automotive industry. The CAN is a low-cost, lightweight network that helps various CAN devices to communicate with one another. It allows electronic control units (ECUs) to have a single CAN interface rather than analog and digital inputs to every device in the system. This decreases overall cost and weight in automobiles. The communication within CAN network is transmitted as broadcast message for which all devices equipped with CAN controller chip can see the message and accept it if the message is destined for it. The CANOpen is basically a software add-on to provide network management function to CAN. Both CAN and CANOpen are used as a fieldbus in embedded solutions.

12.2.3.5   DeviceNet

This is another ODVA supported technology specifically defined for fieldbus use. It uses CAN technology for PHY and CIP for upper layer. The protocol provides open, device-level control and information networking for simple industrial devices. It supports communication between sensors and actuators and higher-level devices such as PLCs and computers. With power and signal in a single cable, it offers simple and cost-effective wiring options. Figure 12.6 shows single cable connectivity that forms a trunkline-dropline topology of DeviceNet fieldbus.

It uses same cable for network communications and DC power up to 24 Vdc, 8 Amps. The DeviceNet operates in a master-slave or a distributed control architecture using peer-to-peer communication, and it supports both I/O and explicit messaging for a single point of connection for configuration and control [9].

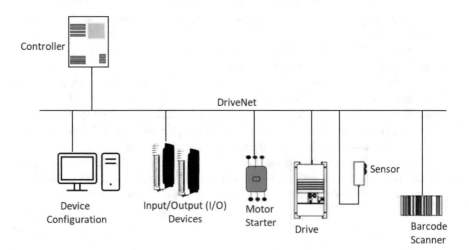

Fig. 12.6 Typical DeviceNet topology

Table 12.2 Industrial applications and corresponding time sync requirements

Application	Time Sync needs
Motion control (newspaper printing)	4 µs (relative time)
Laser cutting and marking machines	<100 ns (relative time)
Electrical substations (differential protection)	10 µs (absolute time)
Electrical grids: WAN protection	1 µs (absolute time)
Drive (GTO, IGBT firing)	1 µs (relative time)

12.3 Synchronization Need for Industrial Networks

As evident from the earlier discussion on various communication protocols and technologies used in industrial networks and fieldbus is that determinism and real time are two key imperatives of industrial networks and for that matter industrial systems interconnect. The determinism creates a lossless network with guarantee on traffic transport without delay while real-time control ensures that endpoints within the networks are precisely synchronized. Much of industrial network deployments require time-critical operations for various industrial systems, e.g., motion controllers, I/O devices, and actuators. If these motion control systems are not properly synchronized, the factory operation will fail and even cause hazardous condition. Table 12.2 gives a glimpse of some of the industrial applications and its corresponding time sync needs.

In this section, we will explore how synchronization is achieved in each industrial network technologies where applicable.

12.3.1 Synchronization in Industrial Ethernet

We discussed about one of the key aspects of industrial network and specifically for IE, is the real-time control requiring IE to facilitate connectivity of real time (RT) systems. The context of "RT" does not necessarily mean faster processing, but rather that a process is dependent on the progression of time for valid execution. Figure 12.7 depicts a typical IE setup with RT nodes. RT systems depend on both validity of data and timeliness for operational efficacy and IE must guarantee the transport of valid data with proper integrity check in a timely manner. Without such operation guarantee RT system will render in failure. There are two types of RT systems: Hard Real Time (HRT) and Soft Real Time (SRT).

Failure of timeliness and determinism in data transport for HRT system connectivity could cause catastrophic failure or even death in some environment such as flight control or train control. While it is not encouraged, network-induced error condition in SRT system connectivity will not result in loss of life or property. IE uses a number of technologies (such as EtherNet/IP, PROFITNET, EtherCAT, Modbus TCP, and POWERLINK) discussed earlier to guarantee both determinism

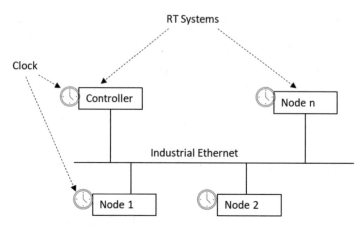

Fig. 12.7 Typical IE setup with RT nodes

and real-time control. Each of these technologies use some specific mechanisms for synchronization to ensure real-time control. Some uses PTP and TSN while other uses proprietary or PTP complaint sync plane integrated within their protocols.

12.3.1.1   EtherNet/IP Sync Plane

EtherNet/IP uses CIP Sync protocol for distributed network wide synchronization. CIP Sync is a protocol defined by ODVA for synchronization in EtherNet/IP. This protocol provides mechanism to synchronize clocks between I/O devices and other automation products with minimal user intervention. CIP sync is considered an extension of CIP and the protocol is IEEE1588 complaint. It allows synchronization accuracy between two devices of fewer than 100 ns. The protocol consists of a time sync object and associated services that allows users to synchronize devices. It provides accurate real-time (Real-World Time) or Coordinated Universal Time (UTC) synchronization of controllers and devices connected over EtherNet/IP networks. The technology supports highly distributed applications that requires timestamping, sequence of events recording, distributed motion control, and increased control coordination. Figure 12.8 depicts a typical CIP Sync setup for EtherNet/IP Network. CIP Sync supports IEEE1588-2008 (PTP Version 2) and the CIP Sync profile specifies the same set of defaults as the Delay Request/Response Default PTP Profile in the 1588 standard.

In most cases, these defaults satisfy industrial application requirements for distributing time throughout the control system. The defaults, such as multicast messaging, sync updates in the one-second range, and the delay request-response path measurement mechanism are all suitable for implementing clocks over industrial networks to meet synchronization requirements [10]. However, it is to be noted that CIP Sync profile currently supports only one PTP clock domain "domain 0," to simplify device and system implementation and deployment.

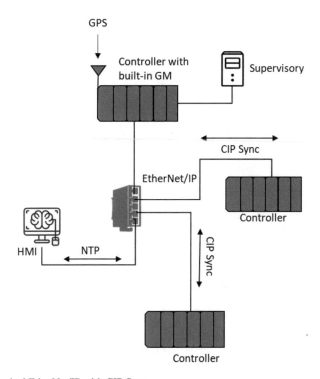

Fig. 12.8 Typical EtherNet/IP with CIP Sync

The PTP defines mechanism to distribute and synchronize time but does not define how to handle perturbations that may occur due to user adjusting master clock or a newly added grandmaster with better clock or grandmaster is temporarily unavailable. To handle this time steps situations and provide stable clock for industrial control systems, CIP Sync defines a clock model as depicted in the diagram below.

The local clock is a frequency disciplined clock that gets tuned each time a PTP Sync/Follow-Up message is received. As shown in Fig. 12.9, local clock is used to schedule all periodic or cyclic operations on the device. A clock offset is also maintained between local time and PTP time. A device will read local time and offset value to get current time.

12.3.1.2   PROFITNET Sync Plane

In our earlier discussion about PROFINET, we understood that the protocol provides capability to build a network with determinism and real-time control for transport. PROFINET defines three categories of devices: Conformance Class A (CC-A), Conformance Class B (CC-B), and Conformance Class C (CC-C). Table 12.3 depicts various types of PROFINET classes and their requirements.

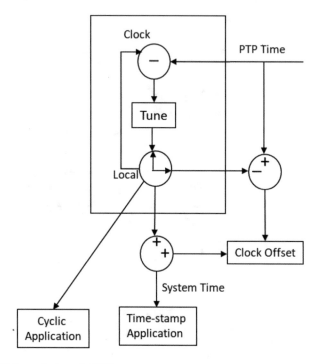

Fig. 12.9 CIP Sync clock model [10]

Table 12.3 PROFINET device conformance requirements

Conformance requirements	CC-A	CC-B	CC-C
Real-time data exchange—Cycle times down to 1 ms	Yes	Yes	Yes
Alarms and diagnostics	Yes	Yes	Yes
Network topology support	Yes	Yes	Yes
SNMP support	N/A	Yes	Yes
Real-time data exchange—Cycle times down to 31.25 μs	N/A	N/A	Yes

Using switches, PROFINET devices both CC-A and CC-B leverage standard Ethernet infrastructure to achieve cycle times as short as 1 ms and jitter of about 10–100 μs. This is the standard "Real Time" (RT) PROFINET communications channel. However, some applications such as closed-loop motion control require both low cycle times and deterministic behavior from their network. For these applications, data should be strictly sequenced and should never be vulnerable to collisions or jitter. To support these applications, PROFINET defines a new MAC-layer extension to Ethernet that allows each switch on an IRT network to provide time slices that turn it into a TDMA (Time Division Multiple Access) medium part of the time and a regular CSMA-CD medium the rest of the time [11].

The IRT networks provide some benefits over the RT networks in deterministic behavior and it can operate with cycle times as low as 31.25 μs (32,000 samples per second). Additionally, IRT has ability to lock the network data exchange to the real-world I/O data process, eliminating aliasing or other sampling artifacts. Figure 12.10 shows typical factory data processing time requirements of the RT processing for factory automation and IRT processing for "motion control."

In a PROFINET based IE network, all switches and devices must be certified based on class requirement discussed above, specifically for IRT switches and devices in the traffic path must conform to CC-C and capable of handling IRT traffic that is time slotted for fast ethernet transport. To achieve IRT communication channel, underlying hardware support is required. Figure 12.11 depicts how time slotted communications happen in PROFINET network for IRT and RT Transport.

Through a clock synchronized synchronization of the involved equipment (network components and PROFINET hardware), a time slot can be defined within the network, by transmitting the important data for the automation task. The communication cycle will be separated into a deterministic part and into an open part. In the deterministic channel, the cyclic clock synchronized real-time telegrams are transported, while standard Ethernet telegrams are transported in the open channel [12].

The clock synchronization is done using PTP protocol. PROFINET defines the PTCP (Precision Transparent Clock Protocol) to share the time reference for IRT network. The PTCP is very similar to PTP version 2. The PTCP is based on an infrastructure composed of cascaded transparent switches (similar to PTP v.2 transparent clock or TC). It uses the peer delay mechanism to measure the port-to-port propagation delay between two ports (i.e., the link delay). It is to be noted that IEEE1588-2019 (PTP version 2.1) supports mapping between PROFINET PTCP and PTP making it easier for seamless integration of PROFINET network with PTP network. Annex H of IEEE1588-2019 defines parameters and other criteria for communication between PROFINET (IEC 61158 Type 10) and PTP version 2.1 devices. Table 12.4 depicts parameter mapping between PROFINET PTCP and PTP as defined in IEEE1588-2019.

The deployment of PROFINET IRT network and PTP network should be done with region isolation using a boundary clock (BC) as shown in Fig. 12.12.

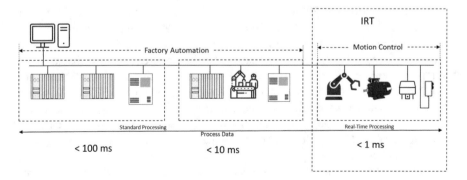

Fig. 12.10 Time critical data processing requirement for factory automation and motion control

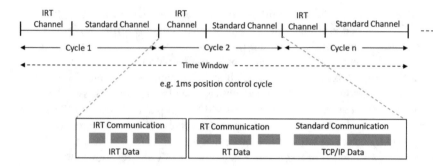

Fig. 12.11 IRT and RT communication channel in PROFINET network

Table 12.4 PTP and PROFINET PTCP parameter mapping as defined in IEEE1588-2019	PROFINET names	PTP names
	SyncPDU	Sync
	FollowUpPDU	Follow_UP
	AnnouncePDU	Announce
	Not used	Delay_Req
	Not used	Delay_resp
	DelayReqPDU	Pdelay_Req
	DelayResPDU	Pdelay_Res
	DelayFuResPDU	Pdelay_Resp_Follow_Up
	Not used	Signaling
	Not used	PTP management

For further detail on various parameters applicable to PTP and PROFINET network integration and implementation in devices, e.g., BC that support PROFINET, please refer to IEEE1588-2019.

12.3.1.3 EtherCAT Sync Plane

Unlike EtherNet/IP and PROFINET, EtherCAT uses Distributed Clock (DC) for a tighter synchronization of <100 ns. The DC refers to a logical network of synchronized distributed local clock in EtherCAT System. The calibration of the clocks in the nodes is completely hardware-based. The time reference is distributed to all slaves by means of the EtherCAT master. For system synchronization, all slaves are synchronized to one reference clock. Typically, the first ESC (EtherCAT Slave Controller) with DC capability after the master within one segment holds the reference time (System Time). This "System Time" is used as the reference clock to synchronize the DC slave clocks of other devices and of the master. The propagation delays, local clock sources drift, and local clock offsets are taken into account for the clock synchronization. With this mechanism, the field device clocks can be precisely adjusted to this reference clock. The resulting jitter in the system is

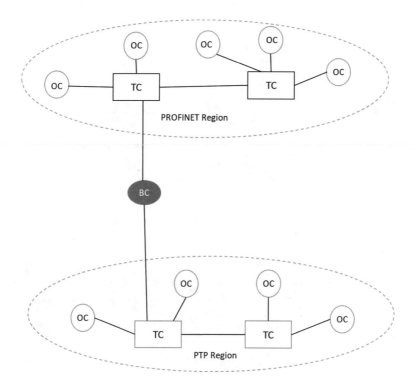

Fig. 12.12 Integration of PROFINET and PTP networks using boundary clock

significantly less than 1 µs. Figure 12.13 depicts typical EtherCAT hardware based distributed clock deployment.

Since the time sent from the reference clock may arrive at the field devices with slight delay, the incurred propagation delay must be measured and compensated for each field device in order to ensure synchronicity and simultaneousness. This delay is measured during network startup or, if desired, even continuously during operation, ensuring that the clocks are simultaneous to within much less than 1 µs of each other. With time information precisely aligned for all nodes in the network, devices can set their output signals simultaneously and affix their input signals with a highly precise timestamp. In motion control environment, velocity measurement is important since it is synchronous with timing. Even very small inaccuracies in the position measurement timing can translate to larger inaccuracies in the calculated velocity, hence periodic position measurement timing is a must. In EtherCAT network, the position measurements are triggered by the precise local clock and not the bus system, leading to much greater accuracy [13].

The use of DC in EtherCAT helps unburden control unit since position measurement is triggered by local clock instead of when the frame is received. Therefore, EtherCAT master only needs to ensure that the EtherCAT frame is sent early enough before the DC signal in the nodes triggers the output.

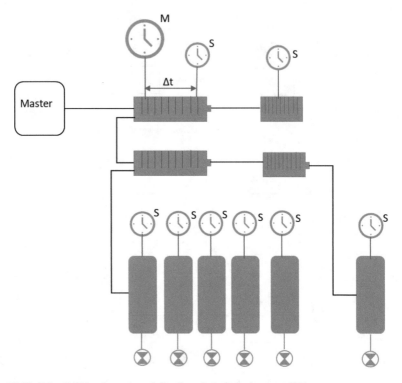

Fig. 12.13 EtherCAT hardware based distributed clock deployment [13]

For synchronization of DC, EtherCAT network uses three distinct methods of synchronization: free running clock, synchronized to output event and synchronized to the sync signal as shown in Fig. 12.14.

The free running mode in which applications are running independently is illustrated in Fig. 12.14a. The "synchronized to output" event is shown in Fig. 12.14b. In this mode, slave application is synchronized to an output event. If no outputs are used, the input event is used for synchronization. The third mode of synchronization is identified in Fig. 12.14c in which application is synchronized to sync signal.

12.3.1.4 Modbus TCP Sync Plane

Unlike EtherCAT, Modbus TCP including Modbus RTU do not have any embedded time synchronization mechanism for high-precision synchronization. Though Modbus TCP allows synchronization of the internal clocks of Modbus TCP capable devices, the accuracy is ±1 s and thus it is not useful for many applications.

Depending upon the device configuration, a device supporting modbus may allow the configuration to use its internal clock or external time synchronization mechanism such as IRIG-B, NTP, or PTP.

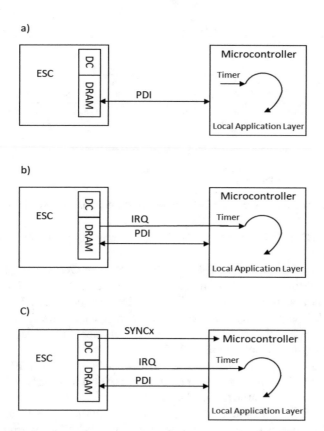

Fig. 12.14 Different synchronization modes in EtherCAT network [14] (**a**) Free running. (**b**) Synchronized to output event. (**c**) Synchronized to the Sync signal

12.3.1.5 POWERLINK Sync Plane

As discussed earlier, POWERLINK is a RTE (Real Time Ethernet) protocol and thus it guarantees isochronous and asynchronous data transfer with dedicated timeslot. POWERLINK also complies with IEC61158 standard for industrial hard real-time protocol. According to IEC61158-2017, POWERLINK provides the following services for real-time communication:

- *Isochronous Data Transfer*: In this operation, one pair of messages per node is delivered at every cycle, or every nth cycle in the case of multiplexed Controller Nodes (CNs).
- *Asynchronous Data Transfer*: One asynchronous message per cycle. It is to be noted that asynchronous data transfer is used for non-time-critical application.
- *Synchronization of all nodes*: At beginning of the isochronous phase, the Master Node (MN) transmits the multicast SoC message very precisely to synchronize all nodes in the network.

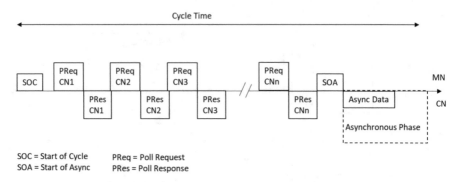

SOC = Start of Cycle   PReq = Poll Request
SOA = Start of Async   PRes = Poll Response

Fig. 12.15 POWERLINK cycle time for isochronous and asynchronous phase of data transfer [15]

The POWERLINK cycles guarantee timeslot for isochronous and asynchronous data transfer.

The cycle is initiated by the MN issuing a Start-of-Cycle (SoC) frame. This frame is used for network synchronization and indicates to each CN to sample its input data as well as to set active the latest received output data. In the isochronous phase, the MN sends PollRequest (PReq) messages, which convey data from MN to polled CN. Each CN then sends its own data to all other nodes via PollResponse (PRes) message as depicted in Fig. 12.15.

The third cycle marks beginning of asynchronous period with SOA (Start of Async) which allows for the transfer of non-time-critical data packets, e.g., parameterization data or IP traffic (Async Data).

The POWERLINK cycle time as shown in the figure above is typically 250 μs and response time is <1 μs. For POWERLINK network, jitter is expected to be <1 μs. According to IEC61158, a hub between MN and CN can introduce a maximum of 70 ns jitter. However, irrespective of topology max jitter limit for the network should be <1 μs. Additionally, IEC61158 recommends the use of PTP boundary clock by POWERLINK routers since PTP is useful for high-precision time synchronization. In this case, PTP time synchronization can be considered as external source and only implemented by the POWERLINK router.

12.3.2   Fieldbus Sync Plane

While fieldbus provides real-time control for various industrial applications, there is no common mechanism of time synchronization among them. Some uses proprietary mechanism or relies on upper layer to provide clock synchronization while other uses external clock source for the devices without any embedded sync mechanism. For example, Modbus RTU does not provide any embedded time sync mechanism. PROFIBUS DP version 2 supports timestamping. For this purpose, a station is defined as the time master to distribute the time within its network. This time master must be a master node and is designated as a class 3 master. In a PROFIBUS DP network, there are three DP masters: Class 1, Class 2, and Class 3. The Class 3 master is responsible for time synchronization. At the beginning, Class 3 master reads its current time and starts an internal timer. As a soon time master sends out time event telegram, the internal timer is stopped. On receipt of TE telegram, the receiving station also starts its own internal timer. The value of this internal timer plus the current value from the TE telegram and the correction value from the counter value (CV) telegram give the time to be set. It is to be noted that this clock synchronization service must always require the support of hardware to start and stop the timers. Currently, not all hardware available on the market can support this service.

On the other hand, synchronization for DeviceNet is supported in PTP version 2.1. IEEE1588-2019 specifies transport of PTP over DeviceNet, messaging structure and timestamping mechanism in Annex F.1. Each PTP instance in DeviceNet has its own multicast address and these addresses are shared in the DeviceNet for PTP communications.

The CANOpen devices use two distinct methods for synchronization: Timestamp Protocol and SYNC Protocol. The timestamp protocol enables the user of CANopen systems to adjust a unique network time. The timestamp is mapped to one single CAN frame with a data length code of 6 byte. These six databytes provide the information "Time of Day." This information is given as milliseconds after midnight (Datatype: Unsigned28) and days since January 1, 1984 (Datatype: Unsigned16). The associated CAN frame has by default the CAN-Identifier 100 h.

The SYNC protocol is transmitted periodically by the SYNC producer. The time period between two consecutive SYNC messages is the communication cycle period. The SYNC message is mapped to a single CAN frame with the identifier 80 h according to the predefined connection set. By default, the SYNC message does not carry any data (DLC = 0). Devices that support CiA 301 Version 4.1 or higher may optionally offer a SYNC message, which provides a 1 byte SYNC counter value. Therefore, synchronous behavior of several devices can be coordinated more comfortably [16].

As for CC-Link, there is inherent time synchronization specified for it. However, a CC-Link family protocol known as CC-LINK IE supports TSN for synchronization. It is to be noted that TSN uses IEEE1588 (PTP) for time synchronization across TSN capable devices.

References

1. Polytron. (2020). *Is Ethernet deterministic? Does it matter?* Polytron.
2. Kim, D., & Tran-Dang, H. (2019). *Industrial sensors and controls in communication networks: From wired technologies to cloud computing and the internet of things.* Springer.
3. HMS. (2020). *Industrial network market shares 2020 according to HMS Networks.* HMS Networks.

4. Profinews. (2020). *PROFINET: The world's leading industrial Ethernet solution.* PI International and PI North America.
5. Ayllon, N. (2020). *What is profinet IRT?* PI North America.
6. Ethercat.org. (2020). *EtherCAT – The Ethernet fieldbus.* EtherCAT Technology Group.
7. EPSG. (2018). *Ethernet POWERLINK communication profile specification version 1.4.0.* EPSG (Ethernet POWERLINK Standardization Group).
8. OpenPOWERLINK. (2021). *Protocol architecture.* openPOWERLINK - An open source POWERLINK protocol stack.
9. ODVA. (2021). *DeviceNet®.* ODVA.
10. Harris, K. (2008). *An application of IEEE 1588 to industrial automation.* IEEE.
11. ProfinetUniversity. (2021). *Isochronous real-time (IRT) communication.* Profinet University.
12. Siemens. (2021). *Why is hardware support required for using clock-synchronized real-time communication (IRT) in PROFINET?* Siemens AG.
13. EtherCAT. (2020). *EtherCAT – The Ethernet fieldbus.* EtherCAT Technology Group.
14. Chen, X., Li, D., Wan, J., & Zhou, N. (2016). A clock synchronization method for EtherCAT master. *Microprocessors and Microsystems, 46*(Pt B), 211–218.
15. openPOWERLINK. (2021). *POWERLINK mechanism.* openPOWERLINK. Available at http://openpowerlink.sourceforge.net/web/POWERLINK/Mechanism.html
16. CiA. (2021). *Special function protocols.* CAN in Automation. Available online at https://www.can-cia.org/can-knowledge/canopen/special-function-protocols/

Index

Printed in the United States
by Baker & Taylor Publisher Services